Kelvin and Stokes

The temperaments of the two were, however, very different. When Kelvin was speaking, Stokes would remain silent until Kelvin seemed at any rate to pause. On the other hand, when Stokes was speaking, Kelvin would butt in after almost every sentence with some idea which had just occurred to him, and which he could not suppress. I once saw a curious reversal of this. They had come together to the Cavendish Laboratory, and I was showing them some experiments I was engaged with at the time, on the electric discharge through gases. I happened to speak about atoms playing a part in one of the effects I was showing, when Kelvin said he did not believe in atoms but only in molecules. This was too much for Stokes. He began at once to give a charmingly clear account of the reasons why atoms as well as molecules must exist. He was so much in earnest that Kelvin for once could not get a word in edgeways: as soon as he started to speak, Stokes raised his hand in a solemn way and, as it were, pushed Kelvin back into his seat.

J J Thomson *Recollections and Reflections*
(London: G Bell and Sons, 1936) pp. 50–1.

Kelvin and Stokes

A Comparative Study in Victorian Physics

David B Wilson

Departments of History and Mechanical Engineering,
Iowa State University

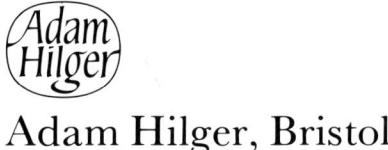

Adam Hilger, Bristol

British Library Cataloguing in Publication Data

Wilson, David B.
    Kelvin and Stokes: a comparative study in
    Victorian physics
    1. Kelvin, William Thomson, Baron
    2. Stokes, Sir George Gabriel, bart.
    I. Title
    530′.092′2   QC16.K3

    ISBN 0-85274-526-5

Consultant Editor: **Professor A J Meadows**,
                 Loughborough University

Published under the Adam Hilger imprint by IOP Publishing Ltd
Techno House, Redcliffe Way, Bristol BS1 6NX, England

Typeset by BC Typesetting
51 School Road, Oldland Common, Bristol BS15 6PJ

Printed in Great Britain by J W Arrowsmith Ltd, Bristol

For Julie

Contents

Preface

Lord Kelvin and Sir G G Stokes have been part of my scholarly life for some time. At Johns Hopkins University my dissertation director, Harry Woolf, first directed me to the Stokes Collection in the Cambridge University Library. Thanks to a Summer Fellowship from the National Science Foundation, I was able to use that collection in writing my dissertation, one chapter of which was the basis for one of my first articles, 'George Gabriel Stokes on Stellar Aberration and the Luminiferous Ether'. In a different form that material has become Chapter 6 in this book. A few years later a NATO Postdoctoral Fellowship in Science enabled me to return to Cambridge for a year to continue work in the Stokes Collection. At the end of that time, the Cambridge University Library, aided by a grant from IBM United Kingdom, hired me as an archivist for their Stokes and Kelvin Collections. It was an invitation for me, as archivist of the Kelvin Collection, to deliver the Kelvin Lecture to the Royal Philosophical Society of Glasgow which first encouraged me to write on Kelvin. Aspects of that lecture, 'Kelvin's Scientific Realism: The Theological Context', appear in Chapter 4. Also arising from that period in the early 1970s are a catalogue of the two collections published by the Cambridge University Library Press and an edition of the correspondence between Kelvin and Stokes, now in press with the Cambridge University Press. Part of my introduction to the latter has been revised and expanded here into Chapter 7. After spending so much time with Kelvin and Stokes I was glad to accept the suggestion of my publishers Adam Hilger that I write this comparative study of their careers and thought.

Many helped prepare me to write the book, and I should like to acknowledge them here. There are four outstanding teachers: Roy Long and Woodrow Pemberton of Benjamin Bosse High School in Evansville, Indiana, and Robert Henry and Lewis Salter of Wabash College. At Johns Hopkins, in addition to Harry Woolf, I benefited greatly from the instruction and advice of Robert Kargon, William Coleman and Maurice Mandelbaum. Duane H D Roller and his views of the history of physics have been sources of inspiration for many years. At Iowa State University the Mechanical Engineering Department and its former chairman, Arthur Bergles, have continually supported my research, financially and otherwise. In Britain, J M A Lenihan has been a limitless font of information and insight into things Scottish, including, of course, Kelvin. Michael Hoskin and David Dewhirst have eased my researches in Cambridge, and Peter Swinbank has done likewise in Glasgow.

I have found participation in two group projects helpful in developing my ideas—one on the Cambridge natural sciences tripos organised by Roy M MacLeod, the other on Cambridge mathematical physics organised by Peter Harman.

I am indebted to numerous libraries and librarians: Cambridge University Library (E B Ceadel, A E B Owen and Peter Gautrey), Cambridge University Archives (Heather Peak, Dorothy Owen and Elizabeth Leedham-Green), Glasgow University Library (N R Thorp, David Weston and J S A Robertson), Glasgow University Archives (Michael Moss and A Wilson), The Royal Society (N H Robinson), Edinburgh University Library (J T D Hall), The Royal Institution (I M McCabe), St Andrews University Library (R N Smart and Christine Gascoigne), Aberdeen University Library (C A McLaren and D B Johnston), The Royal Society of Edinburgh (W H Rutherford and William Duncan, executive secretaries), Trinity College, Cambridge (Trevor Kaye), Pembroke College, Cambridge (W S Hutton), Balliol College, Oxford, the National Library of Scotland, and the Niels Bohr Library of the American Institute of Physics. I am also grateful to Professor E W Laing and Dr R R Whitehead for assistance in using books from Kelvin's library which are housed in the Department of Natural Philosophy of Glasgow University. Quotations from manuscript material in Cambridge University Library are used by permission of the Syndics of Cambridge University Library. In all other cases where I have quoted from manuscript material I am also grateful for permission to do so. In quotations from manuscripts I have used angle brackets ⟨ ⟩ to indicate deleted material.

Though I have drawn on several sources of unpublished material, the principal archives of manuscripts for Kelvin and Stokes are the Kelvin Papers in Glasgow University Library and the Kelvin and Stokes Collections in Cambridge University Library. For descriptions of these archives, see *Kelvin Papers: Index to the Manuscript Collection of William Thomson, Baron Kelvin in Glasgow University Library* (Glasgow: Glasgow University Library, 1977) and D B Wilson (ed) *Catalogue of the Manuscript Collections of Sir George Gabriel Stokes and Sir William*

Thomson, Baron Kelvin of Largs in Cambridge University Library (Cambridge: Cambridge University Library, 1976).

Without financial assistance at key times I could not have written the book. Grants from the Hays Fund of the American Philosophical Society and from the National Science Foundation (SES-8204542) allowed me to use a Faculty Improvement Leave from Iowa State University to spend the spring and summer of 1983 in Cambridge and Scotland. That research was also assisted by a grant from the Royal Philosophical Society of Glasgow. A travel grant from the American Council of Learned Societies enabled me to attend a conference in England in 1984 on Cambridge mathematical physics and to re-examine Kelvin's research notebooks in the Kelvin Collection. Another grant from the NSF (SES-8420679) provided leave for the autumn semester of 1985, during which time I wrote much of the book. Iowa State University has helped with expenses for travel, xeroxing, and microfilming.

I am grateful to *Victorian Studies* for permission to use 'A Physicist's Alternative to Materialism: The Religious Thought of George Gabriel Stokes' as the basis for much of Chapter 4 and to the British Society for the History of Science for permission to use 'George Gabriel Stokes on Stellar Aberration and the Luminiferous Ether' as the basis for Chapter 6.

The manuscript was typed, and retyped, by Audrey Burton, always cheerfully.

Finally, there is the question of what to call William Thomson, Lord Kelvin. My answer was to use Thomson up to 1892, Kelvin afterwards. Hence, the titles of Chapters 7 and 9.

D B W
Ames
August 1986

Sir George Gabriel Stokes (1819–1903)

1819   Born, County Sligo
1835–7   Bristol College, Bristol
1837   Entered Cambridge University
1841   Senior wrangler, first Smith's prizeman
1841–57, 1862–1903   Fellow, Pembroke College
1842   'On the Steady Motion of Incompressible Fluids'
1845   'On the Theories of the Internal Friction of Fluids in Motion…'
1845   'On the Aberration of Light'
1846   'Report on Recent Researches in Hydrodynamics'
1849   'On the Variation of Gravity at the Surface of the Earth'
1849–1903   Lucasian professor of mathematics, Cambridge University
1849   'On the Dynamical Theory of Diffraction'
1851   Elected fellow of the Royal Society of London
1852   'On the Change of Refrangibility of Light'
1852   Rumford medal, Royal Society of London
1854–60   Professor of physics, Government School of Mines, London
1854–85   Secretary, Royal Society of London
1862   'Report on Double Refraction'
1869   President, British Association for the Advancement of Science
1883–5   Burnett lecturer, Aberdeen University
1885–90   President, Royal Society of London
1886–1903   President, Victoria Institute
1889   Knighted
1891–3   Gifford lecturer, Edinburgh University
1893   Copley medal, Royal Society of London
1897   *Conditional Immortality*
1897   Wilde lecture on Röntgen rays
1902–3   Master, Pembroke College

Sir William Thomson, Baron Kelvin of Largs (1824–1907)

1824   Born, Belfast

1834–41   Glasgow University

1841   'On Fourier's Expansions of Functions in Trigonometrical Series'

1841   Entered Cambridge University

1842   'On the Uniform Motion of Heat . . . and Its Connection with the Mathematical Theory of Electricity'

1845   Second wrangler, first Smith's prizeman

1845–52, 1872–1907   Fellow, Peterhouse

1846–52   Editor, *Cambridge and Dublin Mathematical Journal*

1846–99   Professor of natural philosophy, Glasgow University

1847   'On a Mechanical Representation of Electric, Magnetic, and Galvanic Forces'

1851   'On the Dynamical Theory of Heat . . .'

1851   Elected fellow of the Royal Society of London

1854   'Note on the Possible Density of the Luminiferous Medium . . .'

1855   'On the Theory of the Electric Telegraph' (from a letter to Stokes)

1856   Royal medal, Royal Society of London

1856   'Dynamical Illustrations of the Magnetic and the Helicoidal Rotatory Effects of Transparent Bodies on Polarized Light'

1866   Atlantic Cable successful

1867   Knighted

1867   'On Vortex Atoms'

1867   *Treatise on Natural Philosophy*

1870   'On the Size of Atoms'

1871   President, British Association for the Advancement of Science
1873–8, 1886–90, 1895–1907   President, Royal Society of Edinburgh
1883   Copley medal, Royal Society of London
1884   Baltimore lectures
1890–5   President, Royal Society of London
1892   Raised to the peerage
1901   'Aepinus Atomized'

Abbreviations

Br. Assoc. Rep.	*Report of the British Association for the Advancement of Science*
CUL	Cambridge University Library
GUL	Glasgow University Library
Phil. Mag.	*The London, Edinburgh, and Dublin Philosophical Magazine and Journal of Science*
Phil. Trans.	*Philosophical Transactions of the Royal Society of London*
Proc. R. Soc.	*Proceedings of the Royal Society of London*

Stokes

Burnett Lectures	*Burnett Lectures. On Light* 2nd edn (London: Macmillan, 1892)
Memoir	*Memoir and Scientific Correspondence of the Late Sir George Gabriel Stokes* ed Joseph Larmor 2 vols (Cambridge: Cambridge University Press, 1907)
MPP	*Mathematical and Physical Papers* 5 vols, vols IV and V edited by Joseph Larmor (Cambridge: Cambridge University Press, 1880–1905)
Natural Theology	*Natural Theology: The Gifford Lectures* 2 vols (London and Edinburgh: Adam and Charles Black, 1891, 1893)

Kelvin

Baltimore Lectures (1884)	*Notes of Lectures on Molecular Dynamics and the Wave Theory of Light* (Baltimore, 1884)

Baltimore Lectures (1904)	*Baltimore Lectures on Molecular Dynamics and the Wave Theory of Light* (London: C J Clay and Sons, 1904)
MPP	*Mathematical and Physical Papers* 6 vols, vols IV–VI edited by Joseph Larmor (Cambridge: Cambridge University Press, 1882–1911)
PEM	*Reprint of Papers on Electrostatics and Magnetism* 2nd edn (London: Macmillan, 1884)
PLA	*Popular Lectures and Addresses* 3 vols (London: Macmillan, 1891 (2nd edn), 1894, 1891)
Thompson	S P Thompson *The Life of William Thomson, Baron Kelvin of Largs* 2 vols (London: Macmillan, 1910)

1

Stokes and Kelvin: An Introduction

Science in Victorian Britain underwent revolutionary conceptual and institutional changes. Together, thermodynamics and the electromagnetic theory of light, for example, transformed a bundle of only partially linked, largely experimental physical sciences into a coherent, unified, mathematical physics of energy and ether. In the 1890s one could contemplate reducing the phenomena of matter, electricity, magnetism, heat and light to an underlying reality of potential and kinetic energy in an all-pervading ether. The pursuit of scientific research, largely avocational early in the century, was a full-fledged profession by the century's end. Science became important to university curricula, and universities expanded their science faculties. Institutions like the British Association for the Advancement of Science, founded in 1831, and the Royal Society of London, reformed at mid-century, provided organisational support for a growing community of scientists. And that community of late-Victorian scientists resided in a society which, on balance, was much more scientific and less religious than it had been only two or three generations earlier. In sum late-Victorian society endorsed the importance of scientific knowledge and research, and late-Victorian physics affirmed the primary significance of the ideal of unification and the language of mathematics. In these respects, there was an essential *similarity* between late-Victorian Britain and both the 'big science' and the modern physics of the twentieth century. The metamorphosis that created this state of things was the context of the careers of G G Stokes and William Thomson, Lord Kelvin.

Contemporaries, friends, and longtime correspondents, Stokes and Thomson have traditionally been grouped together as major figures in that mid-nineteenth-century 'Cambridge school' of physics which was so important in the revolution of nineteenth-century physics.[1] Not only did this school contribute fundamentally to the new mathematical physical theories, but its members were part of an early-Victorian Cambridge that approximated the professionalisation of the later period. The best students ('wranglers') in the mathematical tripos (an examination in mathematics and mathematical physics) frequently became Fellows of Cambridge colleges and sometimes university professors. They often helped prepare undergraduates for the tripos and even wrote research papers for the two journals published in Cambridge, the *Transactions of the Cambridge Philosophical Society* and the *Cambridge Mathematical Journal*. Stokes and Thomson were high wranglers in 1841 and 1845, respectively. Elected to Fellowships in their colleges, they each coached students for a period before becoming university professors, Stokes at Cambridge and Thomson at Glasgow. As graduates and college Fellows within this enormously prestigious educational system whose members dominated Victorian mathematics and mathematical physics, Stokes and Thomson did hold a Cambridge experience in common.

However, to view them simply as members of this Cambridge school would be to picture them as vastly more similar to one another than they actually were. Indeed, this book's main goal is to provide a comparative study that explores the character of their differences as well as their similarities. Many of the former, the book argues, involved the fact that Thomson attended Glasgow University as well as Cambridge. To begin the task, let us first consider Stokes's and Thomson's careers with respect to four major theoretical developments in nineteenth-century science: the wave theory of light, thermo-dynamics, the electromagnetic theory of light, and the theory of evolution by natural selection.

The wave or undulatory theory of light was the grandest of many early-nineteenth-century accomplishments in the physical sciences. At the turn of the century, Thomas Young published a series of papers proposing that certain optical phenomena could be explained, by analogy with acoustics, on the assumption that light consisted of waves, not bullet-like particles. His experiments on the interference of rays of light seemed analogous to the phenomenon of beats in acoustics. Partially because his theory remained qualitative, Young's ideas were not widely accepted. When the Frenchman Augustin Fresnel began his researches on light in the century's second decade, he had not heard of Young's work. In the French style, Fresnel mounted a sophisticated mathematical analysis of wave motion showing how a wave theory of light could explain diffraction and other optical phenomena. In a well known episode, Fresnel's theory was attacked because it led to the implausible prediction that the shadow behind a small disc would have a bright spot at the centre. When the experiment was performed, however, the bright spot was there. Overcoming great resistance, Fresnel's theory gained general acceptance in France during the 1820s. Light consisted of transverse undulations in an ethereal medium. The Cambridge-educated physicists, John F W Herschel and G B Airy, were prominent in introducing Fresnel's theory into British science in the late 1820s. Airy included it in his course at Cambridge from about 1830, and J D Forbes, under a Cambridge influence, in his natural philosophy course at Edinburgh from the mid-1830s. Though it was Fresnel's work that had shown the wave theory's great power, English writers, especially William Whewell, tended to emphasise Young's importance, interpreting Fresnel as a developer of Young's ideas. There had indeed been some correspondence between Young and Fresnel about the nature of light waves, but it was French physics that produced the modern wave theory, which Cambridge men then helped transmit to the British context. British opposition to the wave theory was strongest in Scotland, especially in the person of the great experimental optician, David Brewster.

Both Stokes and Thomson first studied the wave theory in the late 1830s, Stokes at Cambridge, Thomson at Glasgow. Clear evidence of their own acceptance of the wave theory dates from somewhat later— for Stokes in the mid-1840s, for Thomson in the early 1850s. I know of no reason to suppose that they did not accept the theory earlier, though it is possible that during the 1840s Thomson doubted the existence of an *imponderable* ether, that is one not subject to gravitational attraction. Their respective researches on specific optical problems came at quite different times. After writing about the nature of the ether in the mid-1840s, Stokes published several mathematical studies on optics in the late 1840s and early 1850s. Thomson, though writing about the nature of the ether in the early 1850s, did not attack specific optical problems until the 1880s. His 1884 Baltimore lectures strove to determine all that the wave theory could and could not explain. The lectures were published in 1884, and in a heavily revised edition in 1904. Stokes and Thomson were principal proponents of the concept of an elastic-solid ether, according to which the ether behaved like an ordinary elastic solid in transmitting transverse waves.

Early-nineteenth-century French mathematical physics figured also in the history of energy theory. Emphasising the importance of the heat in steam rather than its pressure, Sadi Carnot published in 1824 a highly original mathematical analysis of the operation of steam engines. Accepting the view that heat consisted of an indestructible kind of matter called caloric, Carnot explained steam engines by analogy with water-power technology. In an overshot water wheel, for example, the effect that water imparts to the wheel is greater the greater the distance through which the water falls while in contact with the wheel. Analogously, in a steam engine, caloric 'falls' from the high temperature of the boiler to the lower temperature of the condenser. The greater the fall in temperature, the greater the mechanical effect derivable from the caloric. Also, the caloric, like water, is conserved—none is used up or lost in the process of producing mechanical effect.

Thomson was aware of Carnot's analysis by 1844. Though he may have doubted the existence of imponderable caloric, he accepted the idea of the conservation of heat. Thus, in 1847, when he heard J P Joule read a paper to the British Association, he was sceptical. Though the eventual collaboration between Thomson and Joule was one of the momentous partnerships in Victorian science, in 1847 they disagreed.[2] Joule claimed to have shown experimentally that heat was not conserved, but was interconvertible at a constant rate of exchange with mechanical effect. Joule argued further that heat consisted of the motion of ordinary matter at the microscopic level. Trying to reconcile Carnot's powerful analysis, based on the conservation of heat, with Joule's experimental results, rejecting the conservation of heat, led Thomson to probably the major publication of his career, 'On the Dynamical Theory of Heat', published in 1851. In that paper, influenced also by the work of his brother, James, the engineer–scientist W J Maquorn Rankine, and the German physicist Rudolf Clausius, Thomson endorsed Joule's dynamical theory of heat and laid down two 'propositions':

> Prop. I. (Joule).—When equal quantities of mechanical effect are produced by any means whatever from purely thermal sources, or lost in purely thermal effects, equal quantities of heat are put out of existence or are generated.
> Prop. II. (Carnot and Clausius).—If an engine be such that, when it is worked backwards, the physical and mechanical agencies in every part of its motions are all reversed, it produces as much mechanical effect as can be produced by any thermodynamic engine, with the same temperatures of source and refrigerator, from a given quantity of heat.[3]

These propositions involved Thomson's ideas of the indestructibility and dissipation not of heat, but energy. The various forms of energy were interconvertible, but the total amount of energy remained constant. The energy did dissipate, becoming increasingly less usable. Carnot rightly emphasised the temperature fall in a steam engine, because the greater the temperature difference, the greater the mechanical effect produced. However, Joule was right that the mechanical effect came from the conversion of an equal amount of heat energy. In Thomson's terminology, the process involved a dissipation of the energy in high-temperature steam to the resultant mechanical energy and the energy in the condensed steam. Generally, there existed 'A Universal Tendency in Nature to the Dissipation of Mechanical Energy.' Among other things, as Thomson declared in 1852, modern physical theory therefore implied that 'within a finite period of time past, the earth must have been, and within a finite period of time to come the earth must again be, unfit for the habitation of man as at present constituted . . .'.[4]

The new energy doctrine was not given immediate universal acceptance. The venerable John Herschel seemed to regard the idea of potential energy as sort of a bookkeeping trick. Obviously, if one defined such potential energy properly, one could always 'find' the energy that had disappeared. In 1871 P G Tait, Edinburgh professor of natural philosophy and co-author with Thomson of a physics textbook based on energy theory, chided members of the British Association's section A (for mathematics and physics). Though the conservation of energy had been generally accepted, Tait complained that 'the Dissipation of Energy is by no means well known, and many of the results of its legitimate application have been received with doubt, sometimes even with attempted ridicule'. If Britons did not accept and develop Thomson's theory of dissipation, 'we shall soon learn its consequences from abroad'.[5]

Though Stokes supported Joule's dynamical theory of heat and the conservation of energy readily enough, he may have been one of the early sceptics regarding dissipation. James has discussed Stokes's employment of the conservation of energy in his paper on fluorescence, read to the Royal Society in May 1852.[6] Stokes invoked the concept, for example, in rejecting a theory that the absorption of light was caused by its annihilation due to interference. 'But we have reason to think', Stokes wrote, 'that the annihilation of work is no less a physical impossibility than its creation, that is, than perpetual motion.'[7] In letters to Thomson, Stokes wrote that the index of friction in pendulum experiments 'is as it were the door through which vis viva passes from the mechanical state (observable motion) to the molecular state (heat)' and remarked that the Frenchmen, Regnault and Foucault, thought certain of their experiments were so novel only because they apparently did not 'perceive the full bearing of the dynamical theory of heat'.[8] The experiments should have been a 'matter of course' to any 'Joulite'.[9] Moreover, the surviving lists of Stokes's lecture topics for his course on physics at the Government School of Mines in the 1850s show that he regularly lectured on the dynamical theory of heat. In 1857, for example: 'Dynamical theory of heat. Rumford's exper[ts]. Joule's researches.'[10] Though this topic could include the conservation of energy, it would not appear to include the dissipation of energy. Indeed, the second law of thermodynamics was not one of the lecture topics in any of the four years from the 1850s for which his lecture notebooks survive. He did deal with other modern research, for example Faraday's work on 'magnetic rotation [of light]', 'line of magnetic force', and 'magnetic field'.[11] Thus, if Stokes did support the idea of the dissipation of energy in the 1850s, he evidently was not an enthusiast for it.

Among the problems to which Thomson applied the concepts of thermodynamics were the ages of the sun and earth. Interested in the distribution of heat within the earth as early as the 1840s, Thomson published something on these questions in the 1850s and presented mature answers in the 1860s. First, contemplating the possible causes of the sun's heat, Thomson eventually concluded that its heat originated from the gravitational potential energy of the parts of matter which coalesced to form it. This idea fitted in with the widely accepted 'nebular hypothesis' that the sun and planets had formed from a condensing cloud of primordial gas. The sun was now an incandescent liquid mass, radiating heat but receiving little or no additional heat. Second, using data from measurements of underground temperatures, Thomson calculated how long it would take a molten earth of uniform temperature to come to its present condition with a particular temperature gradient from a cool surface to a hot centre. Third, Thomson estimated the rate at which the braking effect of tidal action would retard the earth's speed of rotation. The earth's shape corresponded to its speed of rotation when it solidified, which was greater than its present speed. The effect of tidal retardation therefore gave the length of time it would take the speed of the earth's spin to slow from what it was at solidification to its present value. These three lines of argument roughly agreed in limiting the age of the earth to around a hundred million years. Thomson's main target was British uniformitarian geology, which followed Charles Lyell in regarding the earth as indefinitely old. Thomson's estimates, though staunchly opposed by T H Huxley, carried the day in persuading late-Victorian geologists to squeeze their geological chronologies into a shorter period.[12]

Developing ideas of Thomson and Michael Faraday, James Clerk Maxwell, second wrangler in 1854, formulated his electromagnetic theory of light in the 1860s. Thomson published two highly original papers in the 1840s which influenced Maxwell. In 1842 he discussed the similar mathematical structure of static electricity and heat flow. In 1847, impressed by Faraday's discovery (the Faraday effect) that the plane of polarised light could be rotated in a magnetic field, Thomson proposed a unifying 'mechanical representation' for electricity and magnetism. Different activities within a single elastic-solid medium were similar to, and therefore represented, certain electrical and magnetic phenomena. The medium was thus analogous to the luminiferous ether, the role of which Thomson was extending from light to electricity and magnetism. In 1856 Thomson postulated that microscopic magnetic rotations caused the Faraday effect. Shortly after taking his Cambridge degree, Maxwell sought to conceptualise Faraday's 'lines of force' in a Thomsonian manner. He first explored the geometrical similarities between the lines and the flow of an imaginary fluid. In the early 1860s he published his own unifying mechanical representation of electrical and magnetic phenomena, utilising Thomson's concept of microscopic magnetic rotations. The idea was that light consisted of electrical and magnetic rotations that alternately gave rise to one another, thus propagating an electromagnetic wave through the ether at the speed of light. Visible light was composed of such waves within a particular range of wavelengths, but the entire electromagnetic spectrum, according to Maxwell's theory, should also contain waves both shorter and longer than those of light. In 1887 the German physicist, Heinrich Hertz, detected longer waves experimentally, thus gaining acceptance for Maxwell's theory among most of those who did not already accept it. Neither Stokes nor Thomson was among the most enthusiastic supporters of Maxwell's theory.

Stokes always regarded himself as weak in electricity and magnetism. Kelvin's obituary of Stokes cited this area as one which Stokes 'looked upon from outside, scarcely entering it'.[13] It was not that Stokes simply ignored Thomson's areas of research in the 1840s, for much of their correspondence during that decade concerned mathematical analogies that existed among the various areas of physics, especially fluids and heat. Stokes exclaimed in an 1847 letter: 'What an intimate relation there is between the mathematical considerations which are applicable to heat, fluid motion, and attraction!'[14] He also, for example, tried to explain the Faraday effect in letters to Thomson[15] and, in 1854, sent Thomson a detailed discussion of a paper by Faraday, asking Thomson: 'Am I right in attributing the finiteness, and even (for such experiments) considerable magnitude, of the time concerned in the phenomena described by Faraday Phil. Mag. March 1854 p. 197 relative to the charging &c of a long (100 miles) telegraph wire coated with gutta percha and immersed in water to the following two causes?'[16] But Stokes did not publish on electricity and magnetism, nor seek to establish a unified view of physical phenomena. His letters on the Faraday effect were not published until the twentieth century,[17] and it was Thomson, not Stokes, who answered Stokes's question of 1854 with an extended theoretical analysis of submarine telegraphy.

With this background, it is perhaps not surprising that Stokes had little to say about Maxwell's theory. It did not become a part of his lectures at Cambridge on light, for example. However, Stokes was not completely silent on the matter. Discussing the theological importance of the ether in his Burnett lectures for 1883, Stokes cited 'perhaps the most remarkable of all the investigations of the late Professor Clerk Maxwell'—that 'the velocity of propagation of an electrical state' was the same as the velocity of light. This was not support for Maxwell's claim that light consisted of electromagnetic waves, just for the view that purely electrical processes were propagated at the speed of light and, therefore, were probably transmitted by the luminiferous ether. That was sufficient, however, to indicate that the ether seemed to be involved in optical, electrical, and other purposes, 'all bearing so intimately on our well-being'.[18] By 1883, therefore, part of Maxwell's theory was supporting Stokes's view of an ether with the kind of expanded role which, as already mentioned, Thomson had tried to define much earlier. Writing to Thomson in 1884, Stokes entertained the idea that both the old undulatory theory and Maxwell's electromagnetic theory were imperfect approximations to a deeper truth. At one time, he wrote, he had had some success in using the undulatory theory to investigate the propagation of light through crystallised media.

> I mentioned the thing one time to Maxwell at Glenlair, but I never published it, because Maxwell seemed to get out the results from his electromagnetic theory. There is not however I think necessarily opposition between the two; it may be two apparently very different modes of viewing the same thing, but which we might be able to see came to the same thing if we knew more about the real state of things.[19]

In 1890, three years after Hertz's famous experiments concerning Maxwell's theory, Stokes presented the theory more positively. In awarding the Royal Society's Rumford medal to Hertz, Stokes spoke of Maxwell's theoretical value for the speed of an electromagnetic disturbance as itself 'strong reason for believing that light is an electromagnetic phenomenon'. Stokes explained that 'Hertz was the first to detect electro-magnetic waves in free space. . . . These important researches contribute powerfully to the inducements we have to refer the phenomena of light and electricity to a common cause, different as hitherto their manifestations have been.'[20] In his Gifford lectures for 1893, once again extolling the many functions performed by the ether, Stokes declared that Maxwell's theory 'has quite recently received a remarkable confirmation through the investigations of Professor Hertz'.[21] Thus, without giving Maxwell's theory the detailed attention that Thomson did, Stokes apparently had regarded it for some time as an important, though limited, statement about nature's reality and took Hertz's experiments as confirmation of its actual truth.

In contrast to Stokes, Thomson worried over the several aspects of Maxwell's theory for years. His own copy of Maxwell's two-volume *Treatise on Electricity and Magnetism*, for example, contains many marginal notations, most with specific dates which range from 1883 to 1905. Thus, in the margin of Maxwell's §792 on radiation pressure: 'wholly unacceptable K Oct 16 1905' and 'No dynamical proof of this K Oct 16, 1905.'[22] These annotations no doubt will eventually become part of a definitive study of Thomson's on-going struggle with Maxwell's theory. Knudsen has gone some way towards that study in recent publications. He has identified points in Maxwell's theory that Thomson regarded as wrong or deficient: its concept of transverse electrical waves, its idea of displacement currents producing magnetic effects, its imperfect dynamical foundation, and its lack of an explanation of interactions between ether and matter. On the other hand, Thomson agreed with Maxwell that magnetic force resulted from microscopic rotations and that electric effects were propagated in ether at the speed of light. In agreeing on this last point, Thomson, as Knudsen says, 'accepted the general moral of Maxwell's result, but rejected the specific theory behind it'. Whereas younger British physicists tried to improve Maxwell's theory by 'modifying or extending' it, 'Thomson was alone in rejecting much of Maxwell's fundamental doctrine and advocating a return to the old elastic theory of light as the foundation for optical theory.'[23] Following his own train of electromagnetic thoughts, Thomson evaluated Maxwell's specific ideas largely on the basis of how closely they matched his own. It is one of the great ironies in the history of physics that it was Thomson's physical insight that led Maxwell to his revolutionary theory, much of which Thomson himself could not accept.

Charles Darwin's *The Origin of Species* (1859) prompted a bit of discussion. Two of the most famous scenes in that drama occurred at meetings of the British Association. In 1860, Huxley took on Bishop Wilberforce proclaiming that he would rather be descended from an ape than from one who, like Wilberforce, used his abilities to obscure the truth.[24] In 1874 John Tyndall, Huxley's partner in disparaging Christianity, delivered his Belfast address as president of the Association, marshalling energy conservation and Darwinian evolution into an argument for what looked to most like materialism.[25] Though more controversial than its predecessors, Tyndall's address followed something of a recent tradition in which presidents, whether biologists or not, commented on matters of evolution, theology, and/or materialism.

For example, Lord Wrottesly spoke in 1860 of coming 'nearer to our God' and the engineer, W G Fairbairn, in 1861 of contemplating 'the works of an Almighty Power'. In 1863, the engineer, W G Armstrong, thought that Darwinian evolution had excited 'more enthusiasm in its favour than is consistent with that dispassionate spirit which it is so necessary to preserve in the pursuit of truth'. In 1864, Lyell, still reluctant to support Darwin, emphasised how fragmentary the fossil record was. The next year another geologist, John Phillips, imagined the moment when 'in the fulness of time it pleased the Giver of all good to place man upon the Earth, and bid him look up to Heaven'. The physical scientist, W R Grove, spoke in 1866 on continuity in nature, stating that the more living varieties and intermediate fossil species were found 'the stronger becomes the argument for transmutation and the weaker that for successive creations'. There was no presidential address in 1867, but in 1868 the botanist, J D Hooker, supported Darwin's theory much more strongly than Grove had. The overall controversy even invaded section A that year as Tyndall used his presidential address to the section to outline a limited materialism which concentrated on the correlation between thoughts and physical states of the brain. For many, the materialism was not limited enough.[26] After this decade of discussion, Stokes and Thomson were undoubtedly expected to comment on these topics in their presidential addresses. And they did; Stokes in 1869 making one of his many public statements on such matters, Thomson in 1871 one of his few.

Stokes, no doubt responding to Tyndall's address the year before, rejected the idea that life could be reduced to physics and chemistry. Though we should encourage the study of 'the links in the chain of secondary causation' insofar as they govern the phenomena of life, we should not assume that there exists a continuous series of such links sufficient to explain life itself.

> Let us fearlessly trace the dependence of link on link as far as it may be given us to trace it, but let us take heed that in thus studying second causes we forget not the First Cause, nor shut our eyes to the wonderful proofs of design which, in the study of organized beings especially, meet us at every turn.

Furthermore, Stokes argued that 'when from the phenomena of life we pass on to those of mind, we enter a region still more profoundly mysterious'. These phenomena may transcend those of 'mere life' as much as those of life do physics and chemistry.[27]

Just as Stokes had followed one of Darwin's supporters, Hooker, in the presidency, Thomson followed another, Huxley. Though Huxley concentrated on Tyndall's and Pasteur's researches showing that the spontaneous generation of life did not occur in today's world, he also declared that in the quite different physical and chemical environment of earth's early history he would expect life to have evolved from non-life.[28] In his address, Thomson also broached the issue of the origin of life on earth, presenting his meteoric hypothesis, which is discussed in Chapter 4. Though Thomson seemed reluctantly to allow for the possibility of evolution, he soundly criticised Darwinian natural selection as removing God's design from the world. The Darwinian sentiments that Thomson did like were contained in Darwin's 'entangled bank' conclusion to the *Origin*, some of which Thomson quoted:

> There is grandeur in this view of life with its several powers, having been originally breathed by the Creator into a few forms or into one; and that, whilst this planet has gone cycling on according to the fixed law of gravity, from so simple a beginning endless forms, most beautiful and most wonderful, have been and are being evolved.[29]

Though embedded in a series of addresses following publication of Darwin's *Origin*, Stokes's and Thomson's addresses also reflected their lifelong concerns for theological matters. Over the decades, there was a constancy to their views. As the Darwin–Tyndall–Huxley viewpoint gained ground, Stokes and Thomson resisted, Stokes from a more conservative religious perspective than Thomson. Their presidential addresses to the British Association adumbrate this book's consideration of their philosophical and theological views.

In addition to indicating the overall scientific background to Stokes's and Thomson's careers, this discussion begins to delineate differences between them. Stokes's early optical research versus Thomson's in heat, electricity, and magnetism; Thomson's role as a founder of the science of thermodynamics versus Stokes's in its reception; Thomson's detailed rejection of much of Maxwell's theory versus Stokes's general acceptance of it; Stokes's conservative Christianity versus Thomson's somewhat more liberal views. There were two other contrasts between Stokes and Thomson which would be well to mention here since the book does not concentrate on them.

First, Stokes devoted much more time to the administration of science than did Thomson. It was not that Thomson held no administrative positions. He was, for example, editor of the *Cambridge and Dublin Mathematical Journal* for several years and president of the Royal Societies of Edinburgh and London for several more. But these duties did not consume Thomson's career as they did Stokes's. As secretary, president, and council member, Stokes conscientiously attended to the business of the Royal Society of London, including editing the *Philosophical Transactions* for three decades. He even agreed to represent Cambridge University in Parliament from 1887 to 1891, attending regularly but evidently speaking not at all. Many, including

Thomson, regretted that Stokes had accepted great administrative burden. Thomson made the point in correspondence with Stokes,[30] but P G Tait published his views in *Nature*. Criticising the 'short-sighted', 'mole-eyed' British state, he remonstrated:

> What a comment on things as they are is furnished by the spectacle of genius like that of Stokes' wasted on the drudgery of Secretary to the Commissioners for the University of Cambridge; or of a Lecturer in the School of Mines; or the exhausting labour and totally inadequate remuneration of a Secretary to the Royal Society![31]

Stokes, however, appeared quite comfortable with his role. In part, he had accepted the secretaryship, as well as his professorship at the Government School of Mines, to supplement his small income as Lucasian professor. But Stokes may also have seen the 1850s as a period of transition away from physical research, a transition that included marriage in 1857. The 37-year-old Stokes wrote to his fiancée early in 1857 that he had been 'on the point of sinking into an old bachelor' but that now he 'felt that perhaps my marriage with you would be even the turning-point of my salvation'.[32] 'You are quite right in saying that it is well not to go brooding over one's own thoughts and feelings', he wrote to her a few days later, 'and in a family that is easy, but *you* don't know what it is to live utterly alone.'[33] And a couple of months later: 'I too feel that I have been thinking too much of late, but in a different way, my head running on divergent series, the discontinuity of arbitrary constants, etc., etc. . . . I often thought you would do me good by keeping me from being too engrossed by those things.'[34] When it appeared that she might cancel the wedding at the last minute because of his coldness. 'Then it is right that you should even now draw back, nor heed though I should go to the grave a thinking machine unenlivened and uncheered and unwarmed by the happiness of domestic affection.'[35] While one should avoid reading too much into these remarkably candid letters, they more than hint that Stokes himself may have welcomed what others regretted—his abandonment of the lonely rigours of mathematical physics for domestic life and the collegiality of scientific administration.[36]

Second, unlike Stokes, Thomson had virtually an entire career as an engineer, as well as one as a physicist. Stokes did do some practical work, publishing, for example, a 'Discussion of a Differential Equation Relating to the Breaking of Railway Bridges' in 1849 and later serving on a committee to evaluate lighthouse illumination. But Thomson was one of the great engineers of an age of engineering. His decade-long work on the Atlantic Cable led to its success in 1866 and his own knighthood in 1867. His interest in submarine-cable problems aroused by Stokes's letter in 1854, Thomson extended his previous research in mathematical physics to explain the way a signal is transmitted through a long underwater cable. With the help of his students at Glasgow, he investigated the conductivity of commercial copper, not something well understood at the time but crucial to the operation of a cable long enough to span the Atlantic. His highly sensitive 'mirror galvanometer' enabled the detection of weak signals. Thomson's biographer, the electrical engineer, S P Thompson, approvingly quoted the *Glasgow Herald's* proud assessment:

> Is Professor Thomson, the distinguished electrician, without whose inspiring genius this great business had not been so easily achieved, not a Glasgow man? And were the principal electrical instruments employed in testing and working the cable not manufactured by Mr. [James] White, the optician of this city, though under Professor Thomson's directions?[37]

Turning a few years later to problems associated with his love of sailing, Thomson invented a compass, a sounding machine, and machines for analysing and predicting tides. Thomson's various inventions were manufactured commercially by James White's firm, which was renamed Kelvin and James White Ltd, in 1901. Glasgow was a more practical city than Cambridge, and Thomson was a more practical man than Stokes. The great success of his inventions made Thomson a man of wealth, still another difference between him and Stokes.

Within the overall context described here but without pretending to include every possible topic, the following chapters explore several key aspects of Thomson's and Stokes's careers. The chapters divide naturally into four pairs. Chapters 2 and 3 concern university education, dealing with Stokes and Thomson first as students and then as professors. Whereas Stokes was both undergraduate and professor at Cambridge, Thomson spent his earlier undergraduate years at Glasgow, where he also became professor of natural philosophy. Many of the differences they experienced as students thus naturally carried over to the classes they taught. Chapters 4 and 5 discuss broad conceptual issues of theology and scientific methodology. They describe Stokes's and Thomson's differing theological views while arguing that they shared a theological basis for their 'cautious realism' regarding scientific knowledge. If one bears in mind that both Stokes and Thomson believed in *miracles*, one will be prepared to appreciate the role of theology in their thinking. Chapters 6 and 7 analyse the similarities and differences in their approaches to the ether. Along with energy this was one of the two fundamental concepts of mid- and late-Victorian physics. Their differing educations helped bring them to the ether from the perspectives of different branches of physics and that, in turn, had consequences for their respective concepts of the ether. Chapters 8 and 9 turn to their later careers. Emphasising Stokes's correspondence with William Crookes on cathode rays, Chapter 8 addresses the question of Stokes's pervasive, behind-the-scenes influence on physical research in Britain. Chapter 9 discusses Kelvin's much-maligned writings on radioactivity from his own viewpoint, not, as is usually done, that of his opponents. Devoting separate chapters to Stokes and Kelvin in this way underscores the great disparity in their later careers that has already been noted. To some degree, that disparity was rooted in conflicting trends within their early-Victorian background. Indeed, their initial similarities and especially differences, insofar as they persisted through their careers, provide the unifying theme for the chapters of this book, helping to link discussions of cathode rays and radioactivity to physics education in the 1830s and 1840s.

1   See, for example, E Whittaker 1960 *A History of the Theories of Aether and Electricity* 2 vols (New York: Harper and Brothers) I 153.
2   Joule was one of Thomson's principal scientific correspondents. They published jointly a series of papers from 1852 to 1862 on the thermal effects of fluids in motion.
3   Thomson 'On the Dynamical Theory of Heat, with Numerical Results Deduced from Mr Joule's Equivalent of a Thermal Unit, and M. Regnault's Observations on Steam' *MPP* I 178.
4   Thomson 'On a Universal Tendency in Nature to the Dissipation of Mechanical Energy' *MPP* I 514.
5   P G Tait 1871 Presidential address to section A *Br. Assoc. Rep.* pt II pp. 4–5.
6   F A J L James 1983 'The Conservation of Energy, Theories of Absorption and Resonating Molecules, 1851–1854: G. G. Stokes, A. J. Ångström and W. Thomson' *Not. Rec. R. Soc.* **38** 90–2.
7   Stokes 'On the Change of Refrangibility of Light' *MPP* III 397.
8   Stokes to Thomson, 23 May 1854 *CUL Kelvin Collection* S372.
9   Stokes to Thomson, 10 November 1855 *CUL Kelvin Collection* S380.
10   Stokes Lecture notebook for 1856–7, lecture 35 *CUL Stokes Collection* NB7.
11   Stokes Lecture notebook for 1855–6, lectures 49–50 *CUL Stokes Collection* NB5. The other two lecture notebooks are NB4 for 1854–5 and NB8 for 1858–9.
12   See Joe D Burchfield 1975 *Lord Kelvin and the Age of the Earth* (New York: Science History Publications).

13 Kelvin 'The Scientific Work of Sir George Stokes' *MPP* VI 399.

14 Stokes to Thomson, 10 April 1847 *CUL Kelvin Collection* S327.

15 Stokes to Thomson, 12 and 13 December 1848 *CUL Kelvin Collection* S338, S339.

16 Stokes to Thomson, 16 October 1854 *CUL Kelvin Collection* S374.

17 J Larmor 1924 'An Early Formulation by Stokes of the Theories of the Rotatory Polarizations of Light' *Proc. Camb. Phil. Soc.* **22** 76–81.

18 Stokes *Burnett Lectures* p. 17.

19 Stokes to Thomson, 23 April 1884 *CUL Kelvin Collection* S457.

20 Stokes 1890 President's address *Proc. R. Soc.* **48** 473.

21 Stokes *Natural Theology* II 13.

22 Kelvin's copy of Maxwell's *Treatise* is in the Kelvin Library in the Department of Natural Philosophy, Glasgow University. The first of these two comments may refer to Maxwell's §643, which Maxwell cites in §792. Kelvin has underlined the citation. The second comment is opposite Maxwell's paragraph which reads: 'Hence in a medium in which waves are propagated there is a pressure in the direction normal to the waves, and numerically equal to the energy in unit of volume.'

23 O Knudsen 1985 'Mathematics and Physical Reality in William Thomson's Electromagnetic Theory' in P M Harman (ed) *Wranglers and Physicists: Studies on Cambridge Physics in the Nineteenth Century* (Manchester: Manchester University Press), pp. 171–6. See also O Knudsen 1976 'The Faraday Effect and Physical Theory, 1845–1873' *Arch. Hist. Exact Sci.* **15** 271–2.

24 See G Himmelfarb 1968 *Darwin and the Darwinian Revolution* (New York: W W Norton pp. 287–94. There is some debate about what was actually said in the Huxley–Wilberforce exchange. Stokes was in the audience for a Darwinian confrontation between Huxley and the anatomist, Richard Owen, two days earlier. He wrote to his wife 'There was a paper by Dr Daubeny . . . which excited a great deal of interest. The room was filled as full as it would hold, and the paper produced discussion in which Owen and Huxley took part.' (Stokes to M S Stokes, 28 June 1860 *Memoir* I 74.)

25 J Tyndall 1874 Presidential address *Br. Assoc. Rep.* pp. lxvi–xcvii.

26 For these respective presidential statements see *Br. Assoc. Rep.* 1860 p. lxxv; 1861, p. liii; 1863 p. lxiii; 1864 p. lxxv; 1865 p. lix; 1866 p. lxxii; 1868 pp. lxvi–lxxii; 1868 pt II pp. 4–6.

27 Stokes 1869 Presidential address *Br. Assoc. Rep.* pp. civ–cv.

28 T H Huxley 1870 Presidential address *Br. Assoc. Rep.* pp. lxxxiii–lxxxiv.

29 Thomson 1871 Presidential address *Br. Assoc. Rep.* p. cv.

30 Thomson to Stokes, 21 December 1859, 30 November 1884 *CUL Stokes Collection* K109, K274.

31 P G Tait 'Scientific Worthies. V.—George Gabriel Stokes' *Nature* **12** (15 July 1875) 201–2. Stokes had written to Thomson in 1859 that 'I have another iron in the fire now: I have just been appointed an additional secretary of the Cambridge University Commission.' (Stokes to Thomson, 12 February 1859 *GUL Kelvin Papers* S81.)

32 Stokes to M S Robinson, 24 January 1857 *Memoir* I 53.

33   Stokes to M S Robinson, 27 January 1857 *Memoir* I 53.
34   Stokes to M S Robinson, 31 March 1857 *Memoir* I 63.
35   Stokes to M S Robinson, 18 June 1857 *Memoir* I 71.
36   Serving 39 years on the council of the Royal Society, from 1853 to 1892, Stokes participated in governing it longer than anyone else of the time. Things did not always go smoothly, though, as the likes of Huxley opposed his becoming president in 1878 and 1883 before he was elected in 1885, as Huxley's successor. (R Barton ''An Influential Set of Chaps': The X Club and Royal Society Politics 1864–85', unpublished. I am grateful to Dr Barton for allowing me to read and cite her paper.) In 1887, Huxley strongly disapproved of Stokes, as president of the Royal Society, also becoming a member of parliament and holding the presidency of the Victoria Institute. He seemed to Huxley to be linking the Society to politics and conservative religion. When Huxley thus criticised Stokes in *Nature*, Thomson came to his defence, writing to Stokes: 'We were *very much* displeased with that article in Nature. I think on the contrary, your agreeing to be member was most patriotic and public-spirited. Personally, and for the sake of science I can't help feeling sorry that it will take up so much of your time.' (Thomson to Stokes, 23 November 1887 *CUL Stokes Collection* K281. See also *Memoir* I 102, 177–9.)
37   *Thompson* I 493. For an account of the collaboration between Thompson and Kelvin on the biography see J S Thompson and H G Thompson 1920 *Silvanus Phillips Thompson, D.Sc., LL.D., F.R.S.: His Life and Letters* (London: T Fisher Unwin) pp 278–95. Thompson interviewed Kelvin several times during the last year and a half of his life, thus providing a source for otherwise undocumented information in the biography.

2

The Early-Victorian Background

Awareness of the geographical distribution of scientific ideas is one of the more important insights that historians of science have to contribute to an understanding of science and its history. Science is not *simply* international, or at least it has not been. One need only remember the sharp division between British Newtonians and continental Cartesians in the early eighteenth century or recall the great successes of gravitational theory in the hands of *French* mathematical astronomers later in that century. The nineteenth century included the contrast between continental action-at-a-distance concepts and British field theory as well as the disagreement between British and German physicists over cathode rays (discussed in Chapter 8). Though scientific thought may be more international than, say, art or literature, there nevertheless have existed distinct national and regional styles in scientific thought, which individual scientists have of course partaken of. Such differences are the rule rather than the exception, though for any specific period it is historical investigation that must determine exactly how the groups were defined and exactly how their respective concepts differed. Scientists, like others, tend naturally to be more emphatically knowledgeable of conclusions and viewpoints current within their own group. They are not uniformly or equally abreast of thinking going on in all other local groups. However, such groups are neither unchanging nor totally isolated from one another. One problem, therefore, is to understand the character of the interactions among them. That is why the *historical* investigation of scientific thought places so much importance on such questions as how were scientific ideas transmitted from one region to another and who influenced whose thinking? With the careers of Stokes and Thomson

17

in view, this chapter explores one geographical division—that between early-Victorian Glasgow and Cambridge Universities (which involved a second—that between early-nineteenth-century British and French physics).

It has long been recognised that France dominated most aspects of early-nineteenth-century science, including mathematical physics. Indeed, it may even be that *la physique* which was developed in the France of that period constituted the invention of modern physics.[1] Stemming from Newton's *Principia*, the mathematical sciences of mechanics and astronomy had enjoyed a prestigious history. Tied more to Newton's *Opticks*, various experimental physical sciences remained largely non-mathematical until the early nineteenth century, when the French effected a massive mathematisation of the subjects of heat, electricity and magnetism. Most notable was Laplace, who had already so successfully applied Newtonian gravitational theory to problems in celestial mechanics. He sought to reduce physical phenomena to mechanical theories involving particles of matter and the forces acting between them, on the model of physical astronomy. He and his followers like Biot and Poisson achieved great successes with the corpuscular theory of light and with fluid theories of electricity and magnetism, in which the postulated electrical and magnetic fluids consisted of particles acting on one another through short-range forces. In addition, Fourier analysed the conduction of heat, and Ampère followed Oersted's discovery of a connection between electricity and magnetism with his mathematical treatment of electromagnetism. Fresnel contributed his mathematically sophisticated wave theory of light, which went far beyond Young's ideas. Whether all of this should be called the invention of physics or not, it was an enormous *French* accomplishment mostly completed by the 1820s.

In their detailed study, Crosland and Smith trace British natural philosophers' growing familiarity with *la physique* from 1800 to 1840. French ideas were transformed somewhat in their transmission to Britain, and the rate of transmission varied according to subject and to region within Britain. Generally, Crosland and Smith conclude: 'Laplace's physical astronomy was followed by Poisson's theory of electricity, Ampère's theory of electromagnetism, Fresnel's theory of light, and Fourier's heat theory, all having profound implications for British natural philosophy.'[2]

How did the transmission transpire in Glasgow and Cambridge? Crosland and Smith record no publications by those at the University of Glasgow which either summarised or extended French research, though unpublished lecture notes from the 1830s indicate an appreciation of French achievements in mechanics and astronomy. In his lectures as professor of mathematics, for example, James Thomson praised the work in gravitational theory of Clairaut, D'Alembert, Euler, Lagrange and Laplace. On the other hand, those educated in the highly mathematical Cambridge curriculum not only appreciated French work, but were prominent in calling attention to it. John Herschel corresponded with Fresnel and wrote a major encyclopaedia article in 1827 extolling his concepts. In a similar article, William Whewell presented Poisson's electrical theories, and he reported on French accomplishments to the British Association in 1835 in his 'Report on the Recent Progress and Present Condition of the Mathematical Theories of Electricity, Magnetism, and Heat'. G B Airy's *Mathematical Tracts* (1826) presented gravitational theory and, in its second edition in 1831, Fresnel's undulatory theory to Cambridge undergraduates. Cambridge graduates of the 1830s, Philip Kelland and George Green, responded in their research to French ideas about heat and light.

Grattan-Guinness's survey of publications by Cambridge men nicely supplements the study by Crosland and Smith.[3] Looking through three major journals from 1815 to 1840, Grattan-Guinness finds 106 papers written by 23 Cambridge men. The papers represented a substantial assimilation of French mathematical work, but primarily in the areas of mathematics, mechanics, hydrodynamics, planetary theory, astronomy and optics. The fields they comparatively neglected were heat, electricity and magnetism. As we shall see, this pattern of publication by Cambridge men continued through the period of Stokes and Thomson. The pattern speaks of the educational power of the Cambridge system. Cambridge graduates who published did so overwhelmingly in the very subjects of their intense undergraduate reading.

Turning to a comparison of Glasgow and Cambridge, we can already sense similarities and differences.[4] With some variation, it was in the traditional areas of mathematised physical knowledge where French work had its greatest impact at both Glasgow and Cambridge. Mathematical studies of heat, electricity and magnetism had much less influence at both universities. That is not to say, however, that Cambridge men's higher rate of publication was the only difference between Cambridge and Glasgow. In fact, studying natural philosophy at Glasgow was quite different from preparing for the mathematical tripos at Cambridge. The next section discusses the curriculum at Glasgow, as experienced by Thomson, and at Cambridge, as experienced by Thomson and Stokes. The chapter concludes by suggesting how these educational differences helped shape the quite different careers of Stokes and Thomson.

Glasgow and Cambridge

Early-Victorian Glasgow and Cambridge Universities were worlds apart. In the 1830s, 84 per cent of Glasgow students were Scottish, two thirds of the 84 per cent coming from west or central Scotland. Only 6 per cent were English.[5] Cambridge undergraduates were over-whelmingly English. Of the top 200 wranglers from 1831 to 1850, for example, around 90 per cent were English. Only about 4 per cent (including Thomson) were from Scotland and another 4 per cent (including Stokes) were from Ireland.[6] Glasgow students in the Faculty of Arts divided their time more or less evenly among several subjects: Latin, Greek, mathematics, logic, moral philosophy and natural philosophy. Cambridge students vying for an honour's degree trained intensively for the mathematical tripos, which covered mathematics and certain mathematised physical subjects. Most Glasgow students matriculated with rather uncertain secondary educations, and few stayed the whole four-year course to take degrees. Cambridge students tended to have stronger secondary educations, and about 120 per year succeeded in the mathematical tripos to earn honours degrees. Whereas Cambridge distributed glory by awarding high finishes in the tripos competition, Glasgow did so through class prizes for exemplary performance or university prizes for essays on assigned topics. Those who took degrees at Glasgow usually became ministers. Cambridge students entered a greater variety of vocations, but the largest portion went into the church. Hence, Glasgow offered young men from the city and surrounding counties a broad educational preparation for the Presbyterian ministry, while Cambridge offered young Englishmen a mathematical training deemed appropriate for Anglican clergymen. Longstanding geographical and cultural separation had allowed the growth of quite different universities. William Thomson was one of the relatively few to combine the two worlds.

However, Glasgow and Cambridge curricula did express, albeit in different ways, a common view of the physical sciences, one which spread physical theory in a spectrum from mathematical dynamics at one end to experimental chemistry at the other. The closer to dynamics the more advanced the science. As John Herschel explained in his influential *Preliminary Discourse* (1830), proper understanding of the advanced areas of science required 'a degree of knowledge of mathematics and geometry altogether unattainable by the generality of mankind, who have not the leisure, even if they all had the capacity, to enter into such enquiries'. One could see the power of mathematics, for example, in the surprising, but verified, predictions of Fresnel's mathematical theory of light. Fresnel's deductions required a mathematical 'chain of reasoning' so long and complicated 'that no *mere* good common sense, no general tact or ordinary practical reasoning, would afford the slightest chance of threading their mazes'.

Dynamics showed the closest alliance between mathematics and science, and, fortunately, it also described nature's most widespread activity, 'motion, and its communication'. The consequences of the few, distinct axioms of dynamics 'may be pursued, by arguments purely mathematical, to any extent, insomuch that the limit of our knowledge of dynamics is determined only by that of pure mathematics, which is the case in no other branch of physical science'. Moreover, dynamics included 'every question that can arise respecting the motions and rest of the smallest particles of matter, as well as of the largest masses', though investigations of the former were 'extremely intricate' because they involved hypotheses about matter at the unobservable level.

By contrast, chemistry, even though it involved weighing and measuring and possessed the quantitative law of definite proportions, was so undeveloped that Herschel went so far as to list it as a part of natural philosophy 'to which mathematical reasoning has never been at all applied'. Other physical sciences—astronomy, optics, heat, electricity and magnetism—ranged between dynamics and chemistry, both in their mathematical content and in the degree to which they had been made dynamical sciences. Astronomy, of course, was the prime example of a mathematised, dynamical science. Of the others, optics, with Fresnel's undulatory theory, was clearly the most successfully mathematised and came the closest to revealing nature's unseen, dynamical structure. The other physical sciences were much less advanced, but Herschel's *Preliminary Discourse*, as part of the transmission of *la physique* from France to Britain, favourably noted French progress in mathematising these areas and in constructing suggestive dynamical theories. He praised, for example, Ampère's 'electrodynamic theory', Coulomb's and Poisson's mathematisation of electricity, Fourier's works on the conduction of heat, D'Alembert's on hydrodynamics and Laplace's on heat and sound.[7]

In a climate defined by views like Herschel's, it was logical to distinguish between mathematised and less- or non-mathematised physical subjects. Thus, the mathematical tripos emphasised mechanics, astronomy, hydrodynamics and optics, giving much less attention to heat, electricity and magnetism. Similarly, Glasgow's natural philosophy course consisted of two separate parts: the mathematical part devoted to mechanics, astronomy and optics and the experimental part devoted to heat, electricity and magnetism. Conceptually and methodologically, it made sense for the largely *experimental* sciences of heat and electricity to be included in a chemistry course as was done at both Glasgow and Cambridge. Understandably, the subjects of heat, electricity and magnetism first became an important part of the Cambridge examination system, in the 1860s, through the new natural sciences tripos, viewed as essentially an *experimental* sciences tripos in contrast to the mathematical tripos. Heat, electricity and magnetism did not enter the Glasgow and Cambridge curricula in their full mathematical form until well after the student days of Thomson and Stokes. It was Thomson himself who early in his professorship tried to establish *la physique* at Glasgow, but it was not until around 1870 that it found a place in the mathematical tripos.

There are thus distinctions to be borne in mind in examining early-Victorian Glasgow and Cambridge. First, the primacy of mathematics highlighted the distinction between mathematical and experimental physics. Both involved *theory*, but some areas of physics—and some physicists—had reached advanced mathematical levels, some had not. Next, one must distinguish a Frenchman's applying mathematics to a particular area of physics from a Briton's being aware of the fact and from a British university's actually teaching the subject. Obviously, the first two did not necessarily require the third.

Thomson's path through science courses at Glasgow was guided by four men, William Meikleham, J P Nichol, David Thomson and Thomas Thomson. William Meikleham, a much-honoured Glasgow undergraduate, became professor of natural philosophy in 1803 and remained so until his death in 1846. J P Nichol became professor of astronomy in 1836 and shared the natural philosophy lectures with the ailing Meikleham the year Thomson took the course. David Thomson, another much-honoured Glasgow undergraduate, had gone on to Cambridge, where he finished disappointingly low in the mathematical tripos, and returned to Glasgow to substitute for Meikleham from 1840 to 1845. Thomas Thomson was Regius professor of chemistry in the Faculty of Medicine from 1818 to 1852. In the 1838–9 session, William Thomson took both Nichol's astronomy class, in which he won the only class prize awarded, and Thomas Thomson's chemistry class, in which he did not win a prize. However, Thomas Thomson later wrote about William: 'His conduct was exemplary; and the figure which he made, when examined, was highly to his credit, and showed a good knowledge of the principles of science. He also attended the Laboratory, where he was a practical student for some time.'[8] Indeed, Thomas Thomson's laboratory, which gave students a voluntary opportunity for experimental work, was evidently the model for the natural philosophy laboratory later established by William.[9] In the

1839–40 session, William won first prize in the natural philosophy class. What did he learn in these classes?

First, he learned about the importance of mathematics as shown by its successful application to those Newtonian subjects, astronomy and mechanics. Meikleham emphasised the importance of mathematics in the mathematical part of his course. He told his students that Newton, Laplace and Lagrange were the ones who had 'chiefly enriched Natural Philosophy'.[10] To earn a degree with distinction in natural philosophy a student had to be examined over Newton's *Principia* and Laplace's *Mécanique Céleste*. A university prize in 1840 concerned the figure of the earth, and William won it with an essay which drew on Laplace, Clairaut and Poisson as well as the Cambridge mathematicians G B Airy and J H Pratt.[11] Though Nichol's astronomy course was 'popular' and, thus, not highly mathematical, he explained that a 'complete or scientific' course would unavoidably use 'much formal mathematics'.[12] The importance of mathematics to physical studies was reinforced by William's father, who in his calculus textbook urged students to cultivate modern mechanics and physical astronomy, 'as they display, in the most striking manner, the power of the Modern Analysis, and manifest its triumphs in numerous discoveries of the most profound and valuable nature'.[13]

The wave theory of light received mixed reviews from William's professors. Meikleham included optics in the mathematical part of his course as it had been 'brought under the dominion of mathematics'. But that dominion included only the laws of optics, not the nature of light. 'With regard to the nature of light we know nothing.' He did recommend works which discussed the wave theory, most notably those by Airy (who 'has given the best account of the undulatory theory of light') and Herschel.[14] Nichol, by contrast, urged the superiority of the wave theory in explaining polarisation, double refraction, interference and Newton's rings. Fresnel, the mathematician, and David Brewster, the experimentalist, are the only two optical scientists recorded in William's notes. Though the concept of transverse undulations appears to have figured in Nichol's lectures, William's notes do not make the concept explicit and record no discussion of potential problems of envisaging transverse waves. His notes include no discussion of the ether.[15]

William learned a good deal about heat, electricity and magnetism, first in the 1838–9 chemistry course which included electricity and, especially, heat. Thomas Thomson published a substantial *Outline* of the subjects in 1830 with a second edition in 1840. Though the subject of electricity belonged to the professor of natural philosophy at Glasgow, 'the connexion between Electricity and Chemistry is now so close, that it is impossible to be master of the latter without being acquainted at least with the principles of the former'.[16] The subject of heat, Thomas Thomson said, belonged to the professor of chemistry. In fact, Thomas Thomson thought that these subjects, though necessary, took too much time from the chemistry course and that Scottish universities should establish new professorships in *physique* devoted to heat, light, electricity and magnetism.[17] As indicated by his *Outline*, Thomson was well abreast of modern experimental researches. He covered the main work of such as Rumford, Davy, Volta, Oersted, Ampère, Arago and Faraday. He noted, for example, 'the important labours of Faraday, which have thrown so much light upon almost every department of electricity, which have corrected so many mistakes, and determined so many first principles'.[18]

William's study of these subjects continued during the next two sessions in the experimental part of the natural philosophy course. In 1839–40, he heard the lectures of Meikleham and Nichol. Meikleham got far enough in his lectures to indicate the place that heat, electricity and magnetism occupied in the course. Thermometry was evidently to be an important part of the study of heat, while in electricity he called attention to the 'great discoveries' made by Davy and Faraday. Magnetism was 'an agent very similar to electricity'. Indeed, Meikleham thought that 'magnetism and electricity are modifications of the same ⟨substance⟩ agent'.[19] In the event, Nichol gave the lectures on these topics, and he seems to have stressed heat much more than electricity and magnetism. There was some discussion of Oersted's discovery of the connection between electricity and magnetism and Ampère's theory that magnetic effects were caused by electric currents, but heat received more systematic attention. Figure 2.1 shows Nichol's outline of the subject, which included recent experimental work on the undulatory nature of radiant heat. William's notes do not mention Fourier, but he later recalled that Nichol's praise of Fourier's mathematical analysis of heat prompted him to read *The Analytical Theory of Heat* at the end of the 1839–40 session. 'I took Fourier out of the University Library; and in a fortnight I had mastered it—gone right through it.'[20] As in astronomy so also in heat, it seems that, though Nichol neither did research on nor taught high-level mathematical theories, he appreciated their importance. And William's fortnight with Fourier was a decisively formative experience, among other things indicating the level of his mathematical knowledge before he entered Cambridge. Perhaps another formative experience was the

Figure 2.1 From Thomson 'Natural Philosophy Class [1840]' *CUL Kelvin Collection* NB10. (By permission of the Syndics of Cambridge University Library.)

degree altered if the ratio
of the density to the elasticity
is constant. Hence velocity
not affected by the height of
the barometer.

Cladni & Savart.

Heat

Thermometers.

(II) Laws of the Diffus. of Heat

  1. Communication

  2. Radiation.

    1 Permeability of bodies to heat.
    2 Double Refraction and Polarization.

(II) Changes effected by heat in bodies

  1. Dilatation.

  2. Change of state (viz liquid, solid, or gas)

III. Economic Applications of Heat

    Steam Engine.

IV. _____

Measuring of Heat.

Some measurable change of state in bodies, should be adopted for
measuring heat. Dilation is the most advantageous.

Mercury's The advantages of mercury
are that it will bear a large amount
of heat without boiling, and much cold without
freezing.

  To be of any use, the indications of

1840–1 session in natural philosophy. That was David Thomson's first session to replace Meikleham, and it was William who helped him prepare experiments for the lectures.[21] It would appear that many of the experiments repeated Faraday's, for David Thomson, cousin of Faraday, was enthusiastic about his relative's research. William remembered being 'inoculated with Faraday fire' by David.[22]

In summary, from 1838 to 1841, William appears to have become thoroughly familiar with the phenomena of heat, electricity and magnetism. Meikleham evidently encouraged something of a unified view of these branches of natural philosophy. Not only did his professors put him in touch with much modern experimental and mathematical research, but they also articulated the ideal of mathematising physical theory, even though none of them was himself master of that craft.

Finally, at Glasgow William learned Boscovich's theory. This eighteenth-century natural philosopher had claimed that matter consisted of geometrical points which were surrounded by forces, alternating at different distances between repulsion and attraction in accordance with a force curve. At least two versions of the theory influenced nineteenth-century British physical thought. Due to Joseph Priestley, the first version eliminated the material points, thus reducing 'matter' solely to the forces or powers. This version may have been supported by Faraday in the 1840s. The second version, supported by many Scottish natural philosophers, expanded the material point into a very small particle.[23] Exactly which version William was taught is unclear. He may have learned of Boscovich's ideas from Thomas Thomson who employed them as a useful way to explain certain chemical phenomena in his early-century chemistry textbook.[24] As indicated by figure 2.2, William certainly did learn it from Nichol. The lone diagram in William's notes belies the significance that Boscovich's theory held for his early thought. He recalled that Boscovich's theory, as learned from Nichol, was part of his worldview of action-at-a-distance forces with which David Thomson's enthusiasm for Faraday's ideas had to compete.[25] Though Nichol, in an earlier account, emphasised the importance of the forces—as opposed to wholly inert matter—in a theological context that affirmed the reality of the spiritual, it is not clear whether he retained at the centre of the forces either a material point or a particle.[26] William's later thought, as discussed in Chapter 7, agreed with the general Scottish interpretation of Boscovich in envisaging a material particle.

In the Cambridge of the early 1840s, the mathematical tripos came in the January of the student's fourth year and lasted for six days, covering several well defined topics. Shortly afterwards, stronger students could take an even more challenging examination over the same topics to try to win one of the two Smith's prizes awarded each year. In addition to mathematics, the tripos included the successfully mathematised physical sciences: mechanics, hydrodynamics, astronomy and optics. The educational idea was that mathematics trained the mind to think clearly and that mathematised physical subjects showed that clarity was possible outside pure mathematics. With his mind properly trained for orderly thinking, the student could go on to cope with whatever material was involved in his chosen career. As high places in the tripos were taken to foreshadow successful careers, ambitious clergymen- and barristers-to-be spent much time grinding away at hydrodynamics and so on. And students' later successes in their various careers merely confirmed the educational wisdom behind the tripos. The tripos was virtually a national institution, and by finishing near the top a student passed prestige on to his college and secondary school. Though relevant professorial and college lectures existed, private coaches undertook the bulk of the teaching.

Both Stokes and Thomson were coached by William Hopkins, by far the most successful of these early-Victorian coaches. As with the other coaches, instruction was constrained by what the tripos covered. Accordingly, Stokes's notes on his reading with Hopkins have titles like: 'Mechanics, Constrained Motion of a Point', 'Impulsive Force and Impact', 'Motion of Two or Three Bodies Attracting Each Other and Lunar Equalities No. 1', 'Hydrodynamics', 'Sound and Light I', and 'Light II'.[27] Listening to Hopkins's lectures and solving his tripos-like problem papers, Hopkins's students thoroughly explored tripos subjects. Hopkins not only conformed to the system, he defended it. He praised the study of physical astronomy and physical optics, for example, because they revealed mathematics to be 'the only instrument of investigation by which man could possibly have attained to a knowledge of so much of what is perfect and beautiful in the structure of the material universe, and the laws that govern it'.[28]

Indeed, optics was a prominent part of the tripos. The standard textbook was Airy's essay on the undulatory theory in his *Mathematical Tracts*. Airy regarded concepts about the ether's constitution as 'generally probable' but the mathematical theory of light waves as being as 'certainly true' as gravitational theory itself.[29] Hopkins referred his students to Airy's text and followed Airy in treating the ether as constituted of particles bound together by attractive forces. According to the undulatory theory, Hopkins said, 'a perfectly elastic medium pervades all space'.[30] James Challis, Plumian professor of astronomy, also lectured on optics, but his lectures, though well attended by stronger students, were probably more important for their attention to geometrical optics and optical instruments than for his comments on the undulatory theory and the ether. The presence of his lectures, however, contributed to Cambridge's emphasis on the subject. The tripos always contained many questions on geometrical optics which led naturally to the few on physical optics.

Heat, electricity and magnetism were not important tripos subjects. Whewell, who had successfully influenced the mathematical content of the tripos, tried to increase the amount of attention paid to these subjects. In addition to his own published summaries of French mathematical theories, he encouraged Robert Murphy to write a textbook on the mathematical theories of electricity and magnetism for the use of undergraduates.[31] Despite Whewell's endeavours, the tripos system, with its already entrenched mathematical subjects, yielded only slightly to the new subjects. Since examiners set only one or two questions in these areas each year, students could safely ignore them. Hopkins himself did in his coaching. When the newly formed Board of Mathematical Studies formally excluded heat, electricity and magnetism from the tripos at mid-century, they regarded their action as mainly confirming previous practice.

Let us summarise the contrasts between Glasgow and Cambridge. In different ways, both systems resisted *la physique*. For a long time, the Glasgow natural philosophy course had had subjects divided between the mathematical and experimental parts of the course. Introducing

Figure 2.2 (opposite) From Thomson 'Note-book of Natural Philosophy class 1839–40' 14 January 1840 *CUL Kelvin Collection* NB9. In this graph of the force curve for Boscovich's theory, the horizontal axis represents the distance between two 'particles' of matter, and the vertical axis represents the magnitude of the mutual force between them. When the curve is below the horizontal axis, the force is repulsive; when it is above, the force is attractive. Thus, where the curve crosses the horizontal axis, the force is zero. These are points of stable or unstable equilibrium. For the smallest distances, as the distance becomes smaller and smaller the repulsive force becomes larger and larger, with the force curve approaching the vertical axis asymptotically (*B*). At larger, sensible distances (*SY*) the force curve represents gravitational attraction. Thus, the curve should not go below the horizontal axis at *Y*, but should approach the horizontal axis asymptotically. The up-and-down portion of the curve represents alternating repulsive and attractive forces at insensible or molecular dimensions. Nichol evidently introduced Boscovich's ideas in a discussion of capillary attraction, an obvious subject for considering molecular forces. The curve alone does not reveal whether Nichol intended the 'particle' at *A* to be an extended particle surrounded by concentric spheres of attractive and repulsive forces, an unextended material point at the centre of the forces, or simply the forces themselves whose geometrical centre was *A*. (By permission of the Syndics of Cambridge University Library.)

however gave better data and his ~~problem~~ theory was completely verifies.
Till within 50 years a, o, our knowledge of gravitation extended only to solar system, but now it is found that the same holds with the motion of systems of stars, which we see as double stars.

Let ε be the flattening of the earth, q the sec's pend at equator. d be the ~~excess~~ of the length of polar sec pendulum over that at equator.

$$ \varepsilon = 0.00865 - \frac{d}{q} $$

$$ l = q + d \sin^2\varphi = \text{length at lat } \varphi $$

Capillary attraction.

Impossible to exhibit Capillary actions. If a small tube, both ends open be plunged into a fluid, the water rises a considerable height in the ~~tube~~. The rise is inversely prop to diameter of tube. All fluids are influenced, but some as mercury are depressed.

Boscovich's theory.

mathematical theories of heat, electricity and magnetism as a regular part of the course would have necessitated considerable alteration. Such alteration was not likely to be made by the ailing Meikleham, the temporary Nichol, or the substitute David Thomson, even if—as seems not to have been the case—any of the three had been thoroughly conversant with the French mathematical theories. As discussed in the next chapter, it was the *new* professor, William Thomson, who made the change, in line with his own research interests. At Cambridge, the well established routine of coaching and examining for the mathematical tripos would have required enormous revision to accommodate mathematical theories of heat, electricity and magnetism. Moreover, though the mathematical expertise existed at Cambridge for the comprehension of French mathematical theories, Herschel and even Whewell did not regard heat, electricity and magnetism as being as completely understood as were mechanics, astronomy and optics.[32] Practical and intellectual resistance to change thus coincided. At Glasgow, however, at least heat, electricity and magnetism were included in the natural philosophy course, thereby providing a forum for Nichol to extol Fourier and for David Thomson to laud Faraday. On the other hand, if those subjects were absent from Cambridge studies, the curriculum there did investigate both the mathematical and the physical aspects of the wave theory of light much more thoroughly than the Glasgow natural philosophy course appears to have done. As the next section discusses, differences between Glasgow and Cambridge were perpetuated in the careers of Thomson and Stokes.

Stokes and Thomson

Before Stokes entered Cambridge in 1837, he spent two years at Bristol College in Bristol. The college's principal was J H Jerrard, an Irishman who had been a Cambridge undergraduate in the 1820s with Stokes's older brother. It was a natural place for George to enrol. Though the Bristol curriculum included Greek, Latin and English literature, mathematics was clearly prominent. The *Outline* of subjects taught at Bristol declared: 'Experience has proved, that a close application to the exact sciences is the best discipline for the mind. . . . The Mathematics are therefore justly held to be an essential part of every liberal education.' Mathematical studies were to be patterned after those at Cambridge. In his first two years, the student was to 'become acquainted with the Differential and Integral Calculus' and to go on to statics, dynamics, conic sections and the first three sections of

Newton's *Principia*. In his third year, 'he will be occupied with the principles of Hydrostatics and Optics, and with the remainder of the first book of the Principia, as well as with Spherical Trigonometry and Physical Astronomy'.[33]

As Thomson did at Glasgow, Stokes attended a professor's lectures in chemistry at Cambridge in 1844. The course was taught by James Cumming, who, like Thomas Thomson, lectured on heat and electricity. Moreover, Stokes's notes on Cumming's lectures are more thorough than his on those of any other Cambridge professor.[34] Unlike William Thomson, however, Stokes did not encounter chemistry as his first subject in the physical sciences. William was 14 years old in 1838 and, even for the prodigy that he was, that was about two and a half years before his first publication. In 1844, Stokes was 25 and had been immersed in the subjects of the mathematical tripos for nearly a decade, six years as a student, three more in hydrodynamical researches. Though Cumming's course no doubt helped Stokes in the chemical aspects of his later work on fluorescence, it failed to deflect the long-gathering momentum behind his mathematical studies in the 1840s. For Thomson, on the other hand, Thomas Thomson's course was reinforced in the natural philosophy class in each of the next two sessions. Thomson heard a good deal about heat and electricity, Fourier and Faraday. Stokes did not. The implications of their quite different starting points followed them through their careers.

Consider their earliest researches, in the 1840s. Though five years the younger, Thomson was the first to publish. Thomson had learned about Fresnel and the ether, especially at Cambridge, but the research programme that he began before he entered Cambridge gave primacy to Fourier and the Glasgow subjects, not the undulatory theory. His first paper, published in 1841, successfully defended Fourier's mathematics against charges made by the professor of mathematics at Edinburgh University. In 1842 he explored the mathematical similarity of heat flow and static electricity. In 1845 he published a mathematical theory of electricity which considered the ideas of Coulomb and Faraday. In 1847 he suggested a unifying 'mechanical representation' of electrical and magnetic activity. In 1850 he published a mathematical theory of magnetism. In 1851 appeared his paper on the dynamical theory of heat which reconciled the best from Carnot's researches with Joule's. Representing not only the development of Thomson's own basic physical thinking, these papers, as noted in Chapter 1 and as others have explained, formed a research tradition which Maxwell adopted and eventually developed into his electromagnetic theory of light.[35]

Guided by a Cambridge education, Stokes's research matured in a distinctly Cambridge setting and devoted itself predominantly to

Stokes in 1839 at the age of 19. From J Larmor (ed) 1907 *Memoir and Scientific Correspondence of the Late Sir George Gabriel Stokes*, 2 vols (Cambridge: Cambridge University Press) vol. I facing p. 16.

Thomson in 1840 at the age of 16. Pencil drawing by his sister, Elizabeth. Courtesy of the National Portrait Gallery, London.

Cambridge subjects. His first research topic, hydrodynamics, was suggested to him by his Cambridge coach.[36] The study of friction in fluids led to his emphasising the similarity of viscous fluids and elastic solids, which in turn provided insights into the nature of the luminiferous ether. In these studies he found himself continually contradicting a Cambridge professor, James Challis, and their disputes were played out in personal correspondence and journal articles (see Chapter 6). His papers on the figure of the earth grew directly out of discussions of the mathematical tripos in early 1849 by the new Board of Mathematical Studies, of which Stokes was a member.[37] When Stokes became Lucasian professor in 1849, he and Challis agreed that he should take over Challis's optical lectures.[38] His first paper after being elected gave a dynamical theory of diffraction. In connection with his optical lectures, Stokes decided to extend his experimental researches on the subject. There followed papers on the colours of thick plates, the interference of polarised light, and fluorescence, the last of which winning him the Royal Society's Rumford medal in 1852. All along, his researches responded mainly to the interests of a Cambridge audience. Indeed, most of his early papers were presented directly to a Cambridge audience, that of the Cambridge Philosophical Society. Kelvin summarised Stokes's areas of expertise: 'Hydrodynamics, elasticity of solids and fluids, wave-motion in elastic solids and fluids, were all exhaustively treated by his powerful and unerring mathematics.'[39]

The formative power of the Cambridge tripos shaped other research interests besides Stokes's. If they did research, early-Victorian high wranglers tended overwhelmingly to pursue Cambridge subjects. Most well known are the astronomer John Couch Adams (senior wrangler 1843) and the mathematicians J J Sylvester (second wrangler 1837) and Arthur Cayley (senior wrangler 1842). George Green (fourth wrangler 1837) did his work on electricity and magnetism *before* he entered Cambridge at the age of 40, and he afterwards published on sound, light and hydrodynamics. Of the wranglers who had not previously attended a Scottish university, *none* did significant research in electricity and magnetism. The only graduate to publish on heat was Philip Kelland (senior wrangler 1834), whose friendship with Cauchy may well have influenced his decision to pick up on the work of his friend and Fourier.[40] Even two Scottish students who became high wranglers during this period, Archibald Smith of Glasgow and D F Gregory of Edinburgh, published their early work on tripos subjects. Hardly isolated, Stokes in his research was merely doing the customary Cambridge thing, just at a higher level than most.[41]

Glasgow University, by contrast, was part of an environment that was more practical and less mathematical. Following the tradition of the natural philosophy class, Meikleham retained the separate experimental portion of the class as being suitable in 'a town distinguished by its mechanical arts'.[42] Of the prize-winning students in the class during the decade from 1834 to 1843, over half of those who did not become ministers went into business, industry, engineering or medicine.[43] Of the seven most prominent natural-philosophy students (counting James but not William Thomson) during this decade, four made their marks in engineering or shipbuilding.[44] In the early-Victorian industrial city of Glasgow there was a kind of cultural alliance between practical matters and experimental natural philosophy. Thomson embodied the alliance in his career, both in his successes in engineering and in his speculative insights into heat, electricity and magnetism. However, though heat, electricity and magnetism were Glasgow subjects, Thomson was the only Glasgow student of this period to publish substantially on them. Moreover, largely absent from his experimental studies as a student, the French mathematical approach was essential to his later research. Whereas success in the mathematical tripos often preceded mathematical publications in astronomy or optics, success in the Glasgow natural philosophy class did not lead to mathematical studies of heat, electricity and magnetism—except in the case of William Thomson. Nevertheless, while Thomson's research did not represent a customary Glasgow endeavour, with his Glasgow background he investigated the subjects of heat, electricity and magnetism as no Cambridge-only graduate of the period did.

Thomson was much more original than Stokes in transforming aspects of his education into his own distinctive research programme. In the 1840s he was, quite simply, *the* British practitioner of *la physique*. In addition, his attempt in 1847 to provide a mechanical representation of electrical and magnetic activity signalled a more completely unifying vision of physical phenomena than shown by most early-nineteenth-century French physicists.[45] Energy physics, with its unifying implications, fits the vision. And Thomson's search for a unified conception of physical nature owed more to Glasgow natural philosophy than to Cambridge mathematical physics. Moreover, one should not overlook the importance of Boscovich's ideas in Thomson's thinking. Taught to him as an undergraduate at Glasgow, they became part of his own natural philosophy lectures (Chapter 3), appear to have figured in his concept of the ether (Chapter 7), and were fundamental in his opposition to Rutherford regarding atomic structure (Chapter 9). Stokes, on the other hand, did not exhibit Thomson's drive for a unified physics. Though Stokes readily accepted much of energy theory, he had played no role in constructing it. Nor did he even do research in electricity and magnetism. In response to one query from Thomson, Stokes had to write: 'I am however too ignorant of electricity to form a firm notion of what might or might not be expected to be done.'[46] Whereas the Cambridge-educated Stokes took a Cambridge research tradition to new heights of accomplishment, the Glasgow-educated Thomson departed from that tradition, and in doing so profoundly shaped the course of nineteenth-century physics.

The early-Victorian influences extended further than Stokes's and Thomson's early research interests. As discussed in the next chapter, their own professorial lectures sustained the initial differences. Though introducing more mathematical physics, Thomson preserved the overall organisation of the Glasgow natural philosophy class. Theoretically, he emphasised the physics of energy, an emphasis suitable to the broad Glasgow curriculum. Stokes presented more of an ether physics, which was appropriate to the Cambridge emphasis on physical optics. As seen in Chapters 6 and 7, their views of the ether differed. For Stokes, it was primarily an optical ether, which posed mechanical–hydrodynamical problems arising from the interaction of ether and matter. For Thomson, who approached the ether initially through electrical and magnetic considerations, it was a question of speculative insight seeking to comprehend the ether as the unifying material substratum of all physical phenomena.

Finally, it seems to me that their educational divergence even contributed to the divergence of their later careers, which are discussed in Chapters 8 and 9. There Stokes is presented as 'a Victorian correspondent', Kelvin as 'a Victorian natural philosopher'. Whatever other factors contributed to the decline in Stokes's research efforts from the mid-1850s on, surely one factor was the increasing importance of Thomson's and Maxwell's electromagnetic views. Stokes's discomfort with the subject both in his teaching (see Chapter 3) and research must have helped keep him in a career of scientific administration. It was not, of course, that Stokes altogether ceased creative thinking on physical matters (see Chapter 8 for his influential theory of x-rays, for example), but that such thought now usually occurred in reaction to problems posed to him by others. After the mid-1850s, he really had no concerted research programme of his own. Of course, neither was it that Thomson had no administrative duties. However, his were not nearly so burdensome as Stokes's, and Thomson's running battle with Maxwellian theory indicates his continuing theoretical activity. As does his vortex-atom theory of 1867, his Baltimore lectures on optics in 1884, and his intense interest in atomic theory and radioactivity during the last decade of his life. In summary, the regional–institutional split underlying Stokes's and Thomson's different educations eventually helped to create a situation in which their respective careers were accurately described by the terms *correspondent* and *natural philosopher*.

1   For discussion see: D S L Cardwell 1971 *From Watt to Clausius: The Rise of Thermodynamics in the Early Industrial Age* (Ithaca: Cornell University Press) pp. 109–10, 293; T S Kuhn 1976 'Mathematical vs. Experimental Traditions in the Development of Physical Science' *J. Interdisciplinary Hist.* **7** 27–31; S F Cannon 1978 *Science in Culture: The Early Victorian Period* (New York: Dawson and Science History Publications pp. 111–36; P M Harman 1982 *Energy, Force, and Matter: The Conceptual Development of Nineteenth-Century Physics* (Cambridge: Cambridge University Press) pp. 12–44.

2   M Crosland and C Smith 1978 'The Transmission of Physics from France to Britain: 1800–1840' *Hist. Stud. Phys. Sci.* **9** 55.

3   I Grattan-Guinness 1985 'Mathematics and Mathematical Physics from Cambridge, 1815–40: A Survey of the Achievements and of the French Influences' in P M Harman (ed) *Wranglers and Physicists: Studies on Cambridge Mathematical Physics in the Nineteenth Century* (Manchester: Manchester University Press) pp. 84–111.

4   This topic is explored by D B Wilson 1985 'The Educational Matrix: Physics Education at Early-Victorian Cambridge, Edinburgh, and Glasgow Universities' in P M Harman (ed) *Wranglers and Physicists* (note 3) pp. 12–48. On Cambridge, see also D B Wilson 1982 'Experimentalists among the Mathematicians: Physics in the Cambridge Natural Sciences Tripos, 1851–1900' *Hist. Stud. Phys. Sci.* **12** 325–71.

5   W M Mathew 1966 'The Origins and Occupations of Glasgow Students, 1740–1839' *Past and Present* No 33 74–94.

6   These percentages are determined from information in J R Tanner (ed) 1917 *The Historical Register of the University of Cambridge* (Cambridge: Cambridge University Press and J A Venn 1940–54 *Alumni: Cantabrigienses, 1752–1900* 6 vols (Cambridge: Cambridge University Press).

7   J F W Herschel 1966 *A Preliminary Discourse on the Study of Natural Philosophy* (New York: Johnson Reprint Corporation) §§20, 23, 87, 236, 335, 214, 215, 212, 352, 256, 181.

8   T Thomson to Electors to the Professorship of Natural Philosophy in the University of Glasgow, 28 May 1846, *Thompson* I 174.

9   Thomson discussed the chemistry laboratory and his own natural philosophy laboratory in 'The Bangor Laboratories' *PLA* II 479–90. On T Thomson see R Sviedrys 1970 'The Rise of Physical Science at Victorian Cambridge' *Hist. Stud. Phys. Sci.* **2** 130n, and J B Morrell 1969 'Thomas Thomson: Professor of Chemistry and University Reformer' *Br. J. Hist. Sci.* **4** 245–65.

10 Thomson 'Note-book of Natural Philosophy class 1839–40' 9 November 1839 *CUL Kelvin Collection* NB9.

11 *Thompson* I 10. The prize essay is in *CUL Kelvin Collection* NB11.

12 *Outline of the Plan of Instruction Proposed to Be Followed in the Class of Practical Astronomy in the University of Glasgow* (Glasgow: University Press, 1836) pp. v–vi.

13 James Thomson 1831 *An Introduction to the Differential and Integral Calculus* (Belfast: Simms and M'Intyre) p. iii.

14   Thomson 'Note-book of Natural Philosophy class 1839–40' 6 and 9 November 1839.

15   Thomson 'Natural Philosophy Class [1840]' pp. 9–16 from the back of the notebook *CUL Kelvin Collection* NB10.

16   T Thomson 1840 *An Outline of the Sciences of Heat and Electricity* 2nd edn (London: H Ballière) p. viii.

17   *Ibid.* p. 1.

18   *Ibid.* p. 319.

19   Thomson 'Note-book of Natural Philosophy class 1839–40' (note 10) 6 November 1839.

20   *Thompson* I 14.

21   D Thomson to J Thomson Sr 13 July 1846 *CUL Kelvin Collection* Tm26; *Thompson* I 178–9.

22   *Thompson* I 19.

23   See P M Harman, *Energy, Force, and Matter* (note 1) p. 77; F A J L James 1985 '"The Optical Mode of Investigation": Light and Matter in Faraday's Natural Philosophy' in D Gooding and F A J L James (eds) *Faraday Rediscovered: Essays on the Life and Work of Michael Faraday, 1791–1867* (New York: Stockton Press) pp. 142–3; and R Olson 1969 'The Reception of Boscovich's Ideas in Scotland' *Isis* **60** 91–103.

24   L P Williams 1961 'Boscovich and the British Chemists' in L L Whyte (ed) *Roger Joseph Boscovich, S.J., F.R.S., 1711–1787: Studies on His Life and Work on the 250th Anniversary of His Birth* (London: George Allen and Unwin) pp. 160–1.

25   *Thompson* I 20.

26   J P Nichol, Corrected text of twenty lectures delivered in Montrose in 1829 *GUL* MS Gen. 1570.

27   Stokes's notes on his reading with Hopkins are in *CUL Stokes Collection* PA2–PA24. Thomson's less complete set of notes is in *CUL Kelvin Collection* PA11–PA17.

28   W Hopkins 1841 *Remarks on Certain Proposed Regulations Respecting the Study of the University* (Cambridge: J and J J Deighton) p. 10.

29   G B Airy 1842 *Mathematical Tracts on the Lunar and Planetary Theories, the Figure of the Earth, Precession and Nutation, the Calculus of Variations, and the Undulatory Theory of Optics* (Cambridge: J J Deighton) pp. v–vi.

30   Stokes, Notes on Hopkins's lectures on 'Sound and Light I' *CUL Stokes Collection* PA20.

31   On Whewell's influence, see H Becher 1980 'William Whewell and Cambridge Mathematics' *Hist. Stud. Phys. Sci.* **11** 1–48. R. Murphy 1833 *Elementary Principles of the Theories of Electricity, Heat, and Molecular Actions* part I: *On Electricity* (Cambridge: J and J J Deighton). Part I is all that was published. Becher has also surveyed the large amount of research and instruction in the sciences that did exist at Cambridge in the first half of the nineteenth century in 'Voluntary Science in Nineteenth Century Cambridge to the 1850s' *Br. J. Hist. Sci.* **19** (1986) 57–87.

32   On Whewell's views, see Crosland and Smith 'Transmission of Physics' (note 2) pp. 56–7.

33   *Outline of the Plan of Education to Be Pursued in the Bristol College* (Bristol: John Taylor, 1830) pp. 6, 9–10.

34   Stokes 'Notes on Professor Cumming's lectures' *CUL Stokes Collection* PA29–PA31. Stokes also took notes on the lectures of Robert Willis (PA25, PA26), William Hallowes Miller (PA27, PA28), John Stevens Henslow (PA32), Whewell on moral philosophy (PA33), George Peacock (PA34) and James Challis (PA35).

35   See, for example, D M Siegel 1981 'Thomson, Maxwell, and the Universal Ether in Victorian Physics' in G N Cantor and M J S Hodge (eds) *Conceptions of Ether: Studies in the History of Ether Theories* (Cambridge: Cambridge University Press) pp. 239–68 and O Knudsen 1985 'Mathematics and Physical Reality in William Thomson's Electromagnetic Theory' in P M Harman (ed) *Wranglers and Physicists* (note 3) pp. 149–79.

36   *Memoir* I 8.

37   Stokes to Thomson, 29 March 1849 *CUL Kelvin Collection* S346; Thomson to Stokes [31 March 1849] *CUL Stokes Collection* K33A.

38   *Memoir* I 8.

39   Kelvin 'The Scientific Work of Sir George Stokes' *MPP* I 339.

40   Kelland's friendship with Cauchy is mentioned in G Chrystal and P G Tait 1880 Obituary of Kelland *Proc. R. Soc. of Edinb.* **10** 323: 'His earlier mathematical work was very much influenced by his admiration for Fourier and Cauchy. The latter, indeed, was his personal friend.'

41   Early-Victorian Cambridge graduates, their careers, and the topics of their researches are listed in Wilson 'The Educational Matrix' (note 4) pp. 36–7. For a more sociological study of these graduates see H W Becher 1984 'The Social Origins and Post-Graduate Careers of a Cambridge Intellectual Elite, 1830–1860' *Vict. Stud.* **28** 97–127.

42   Meikleham's evidence in *Evidence, Oral and Documentary, Taken and Received by the Commissioners Appointed . . . for Visiting the Universities of Scotland* vol. II: University of Glasgow, Parliamentary Papers **36** (1837) 120.

43   From information contained in W I Addison 1898 *A Roll of the Graduates of the University of Glasgow* (Glasgow: James MacLehose and Sons); Addison 1913 *Matriculation Albums of the University of Glasgow, 1727–1858* (Glasgow); and *Glasgow University Prize and Degree Lists, 1834–1863*.
   Of the 64 prize winners from 1834 to 1843 (not counting Thomson), 31 became clergymen, and I have been unable to identify the careers of 13. Of the remaining 20, 13 went into medicine, engineering, industry, or business. None became fellows of Cambridge colleges. By contrast, of the 99 top wranglers (not counting Stokes and Thomson) during the decade spanning Stokes's and Thomson's undergraduate years (1836–1845), fewer than 10 per cent of the non-clergymen (6 of 69) went into practical careers. Two were merchants, two were actuaries, one was an engineer, one was a naval architect, none was a physician. In addition to the church, Cambridge graduates found careers as college fellows, in the law, and as professors of mathematics or natural philosophy. (Information on Cambridge graduates is in the works cited in note 6.)

44   See Wilson 'The Educational Matrix' (note 4) p. 35.
45   Fresnel's desire to unify physical phenomena in a universal fluid ether
      appears to be the French approach closest to Thomson's in this regard.
      (See P M Harman *Energy, Force, and Matter* (note 1) pp. 21–4.)
46   Stokes to Thomson, 15 November 1851 *CUL Kelvin Collection* S365. See
      also Stokes to Thomson, 16 June 1854 *CUL, Kelvin Collection* S373:
      'I confess I have not been convinced of the possibility of producing
      rotation of heat from materials which have not the property even when
      thermo-electricity is called in, but this is a subject of which I know so
      little that I would not be positive.'

3

Professors

Between them, Stokes and Thomson held professorships for over a century, delivering over 7000 lectures to nearly 8000 undergraduates of two major universities.[1] Their lecturing styles and situations differed greatly. Thomson had assistants to prepare demonstration experiments and, eventually, to give some of the elementary lectures; Stokes did everything himself. Thomson did not lecture from notes; Stokes always drew up lists of topics to be covered. Thomson was known for long digressions, often requiring his assistants to produce demonstration experiments on the spur of the moment; Stokes apparently stuck by his list of topics, except when unexpected sunlight let him perform optical experiments not on the day's agenda. In the larger university, Stokes employed the more modest experimental equipment. Amusing anecdotes, affectionately recorded by former students, surrounded the energetic and outgoing Thomson. The more staid Stokes elicited admiration for clear lectures and experiments that always turned out right. One of the very few students who attended the lectures of both compared them:

> Some years afterwards, the author attended Stokes's lectures on the same subject. His calm reflective style was a great contrast to Thomson's impetuosity, and the primitive apparatus he used, although admirably adapted for its purpose, was very different from the elaborate apparatus Thomson generally employed. Both had singularly winning smiles when lecturing, both were actuated by the same enthusiasm for science, and the relations of both to their students were marked by the most perfect old-world courtesy.[2]

Aspects of Britain's 'expanding' Victorian society formed the general backdrop to Stokes's and Thomson's professorships.[3] The continuing industrialisation of Britain demanded more and more engineers and

41

led to the introduction of engineering education into universities. Similarly, the perceived increase in the practical and intellectual significance of science was accompanied by a growth in the scientific community along with its demands for recognition, desire for organisation and pressure for education in science. There was something of a struggle between scientists and clergymen for intellectual and cultural supremacy. Education generally, at all levels, expanded enormously during the century. Glasgow University's enrolment increased over 100 per cent during the century; Cambridge's doubled from 1840 to 1900. Glasgow awarded four times as many degrees in the 1890s as in the 1840s. Cambridge had one tripos leading to an honours degree in 1840, ten in 1900.[4] At the end of Stokes's and Thomson's professorships, unlike the beginning, a young man could *plan* to study science at university in preparation for a career in the profession of science, probably in education at either the secondary or university level depending on how well he did as an undergraduate.

With such developments in mind, this chapter focuses on Stokes's and Thomson's respective courses at Cambridge and Glasgow. Going beyond the anecdotal, it considers the relevance of each course to its own university curriculum. It looks at the content of the courses and the careers of the students who took them. Stokes and Thomson differed in more than style, for each taught not in a vacuum but in a particular university context. Because their universities were so different from one another, predictably so were Stokes's and Thomson's courses. Stokes had the more famous students; Thomson the greater monopoly on physics education. Hence, for Stokes, a central question is what part did he play in educating Cambridge's well known graduates?; for Thomson, who were his students?

Stokes's Cambridge

Cambridge's curriculum in mathematics and the physical sciences was much different in 1900 from that existing in the undergraduate days of Stokes and Thomson.[5] A student of 1900 could study physics in the natural sciences tripos (NST) as well as in the mathematical tripos (MT) and, indeed, was more likely to do so. The MT included more physical subjects in 1900 than it had in the 1840s, and students were encouraged much more to specialise at the advanced levels of their studies. The teaching staff, already considerable in the 1840s, had greatly increased. The Cavendish professorship of experimental physics, founded in the early 1870s, was accompanied by a seven-man staff at the Cavendish Laboratory by the 1890s. Five university lectureships in pure and applied mathematics were founded in 1884. In the 1890s the Cavendish staff provided most of the instruction for physics in the NST. There was at least one constant factor, however: coaches were still the mainstay of instruction for the MT. E J Routh (senior wrangler in 1854) supplanted William Hopkins as the leading coach in the mid-1850s and trained 27 senior wranglers before retiring in 1888, when other coaches took his place. With an expanded teaching staff and two triposes to choose between (or to combine), the physics undergraduate of the 1890s faced a more varied and specialised curriculum than his counterpart in the 1840s.

Without claiming that Stokes was necessarily a prime instigator of reform, we can still recognise his support of the changes. At Cambridge, curriculum changes, proposed by committees, were subject to university-wide discussion and approval. Stokes was often a member of committees which initiated the changes that were eventually approved. First as an examiner for the MT and then as Lucasian professor, Stokes was a member of the Board of Mathematical Studies that formally defined the content of the MT at mid-century, excluding the subjects of heat, electricity and magnetism. As early as 1860, Stokes thought physical subjects should be better represented at Cambridge and during the 1860s was in correspondence with the Astronomer Royal (and former senior wrangler) G B Airy on the subject. In the late 1860s the Board of Mathematical Studies, acknowledging progress in the mathematisation of heat, electricity and magnetism, recommended their inclusion in the MT. Stokes was also on the *ad hoc* Physical Sciences Syndicate which stated that inclusion of the new subjects required establishment of a professorship of experimental physics and construction of a laboratory under his direction. It was Stokes who explained to the new professor, Maxwell, that his lectures would probably be attended both by NST students and MT students electing physical subjects.[6] Only marginally included in the NST before, physics gained full status in the early 1870s. In 1882 changes in both triposes allowed for undergraduates to study physics in a more advanced and specialised way. Finally, by introducing more mathematics into NST physics around 1890, J J Thomson effected that integrated mathematical–experimental study of physics which had been favoured by his predecessors as Cavendish professor, and by Stokes.

Within all this change, Stokes's course of lectures persisted largely unaltered. Each transition in the overall context, however, appears to have substantially affected the place of Stokes's course in Cambridge undergraduate studies.

Stokes's predecessor as Lucasian professor was Joshua King, senior wrangler in 1819 and president of Queens' College from 1832 until his death in 1857. Elected Lucasian professor in 1839, he did not lecture.[7] The professorial lectures on physical subjects in the MT were given by James Challis. When Stokes replaced King, Challis and Stokes agreed that Challis would lecture on astronomy and that Stokes would take over the subjects of hydrostatics, hydrodynamics and optics.[8]

Elected (unopposed) in October 1849,[9] Stokes gave his first course of lectures the following spring. For the next three and a half decades he gave a course during Easter Term, usually on hydrostatics, hydro-dynamics and optics, but sometimes just on optics. In the mid-1880s he began dividing the subjects between two terms. Each year he listed the topics for every lecture, thus accumulating a mass of notebooks that disclose the content of his course for most of the half century of his professorship.

It stayed essentially the same. His opening lecture usually sum-marised the course, as in 1852:

> Sketch, Foundations of hydrostatics—capillary attraction—gases and vapours—instruments—touch on hydrodynamics—Optics. Not a great while to geometrical optics; physical optics; large part of course, devote considerable time. New subject.[10]

His general strategy was to use the early part of the course to lead into the undulatory theory and the many optical phenomena related to it. The hydrostatic lectures on the basic properties of fluids prepared the way for hydrodynamical equations of fluid motion. Figure 3.1 is Stokes's list of topics at this stage of the course in 1852. The figure illustrates another aspect of the course—Stokes's need to capitalise on sunny days. Here the experiments relate to his discovery of fluor-escence, which he was to present to the Royal Society the following month. After discussing waves in fluids, Stokes turned briefly to sound as a bridge to light. Adopting the method of presentation from William Hopkins, Stokes offered essentially the same set of 'postulates', 'hypotheses', or 'requirements' for the undulatory theory for 50 years. He recorded them in his list of topics for a lecture in 1857: '1. Postu-lates of und^y theory. Ether in free space and pervading solid & fluid bodies (2) light consists in und^{ns} (3) self lum^s bodies in state of vib^n (4) dif^t vel^y in refr^g media (5) production of sensation (6) colour on period (7) Brightness on amplitude 1'. Why ether in bodies.'[11] Again in 1900: 'Requirements of undulatory (1) Ether (2) Agitation of self-luminous (3) Ether within bodies as well as vacuum (4) Velocity within media different from period to period (5) Mode of vision.'[12] According to Stokes, vibrations in ponderable matter gave rise to ethereal undu-lations of light, which were propagated by the ether present both in vacua and in transparent material bodies. Impinging on ponderable matter of the eye, the undulations caused the sensation of vision. Similarities between sound and light helped in understanding the undulatory theory, but there were also significant differences between them. Rayleigh recorded Stokes's comments on the ether in 1864:

> It is necessary to suppose the Lum[iniferous] Ether very rare, otherwise the motions of the planets w^d shew its' [*sic*] existence. This idea of density is quite independent of weight, for the Lum Ether may to [*sic*] not be subject to the influence of gravity. . . . In order to account for the fact of dense substances being transparent, it seems necessary to suppose that the Ether pervades transparent bodies, & that light is propagated within them by its' [*sic*] vibrations.[13]

If the content of Stokes's course remained about the same, its relevance to undergraduates did not. Reflecting other changes at Cambridge, Stokes's course went through various phases.

The key document in understanding the period from 1850 to 1880 is the notebook in which Stokes recorded the fees paid to him by those enrolled.[14] Figures 3.2 and 3.3 portray Stokes's course as revealed by this lecture-fee notebook. Some enrolled in the course after they had already placed as wranglers in the MT, but the great majority were still preparing for the MT, and would become wranglers. Figure 3.3 is limited to these men.

Above all, figure 3.3 shows that Stokes's course was one for the best students until the mid-1870s, with about 80 per cent of the top ten wranglers during that time enrolled. Although the senior wrangler for 1859, James M Wilson, stated that his college tutor advised him not to take Stokes's course because it would not 'pay' for the MT,[15] Wilson was one of only *two* senior wranglers from 1851 to 1876 not to take Stokes's course in preparation for the MT. The other, Charles Niven in 1869, took the course right after the tripos.

Figure 3.3 shows a great increase in the number of lower wranglers from the mid-1860s to the mid-1870s. This would appear to reflect the move during that time to increase the physical content of the MT. The figure also shows a noticeable decline in percentages from the first ten to the second ten to the remainder, thus reinforcing the basic conclusion: the stronger the undergraduate, the more likely he was to take Stokes's course.

Surviving comments indicate that there were great differences between other professorial courses and Stokes's. When J J Thomson attended the 'lectures' of Arthur Cayley, appointed the first Sadlerian professor of mathematics in 1863, there were only three in the course, with Thomson as the only undergraduate. Cayley sat at a table writing on sheets of paper while the students sat around him. As the junior

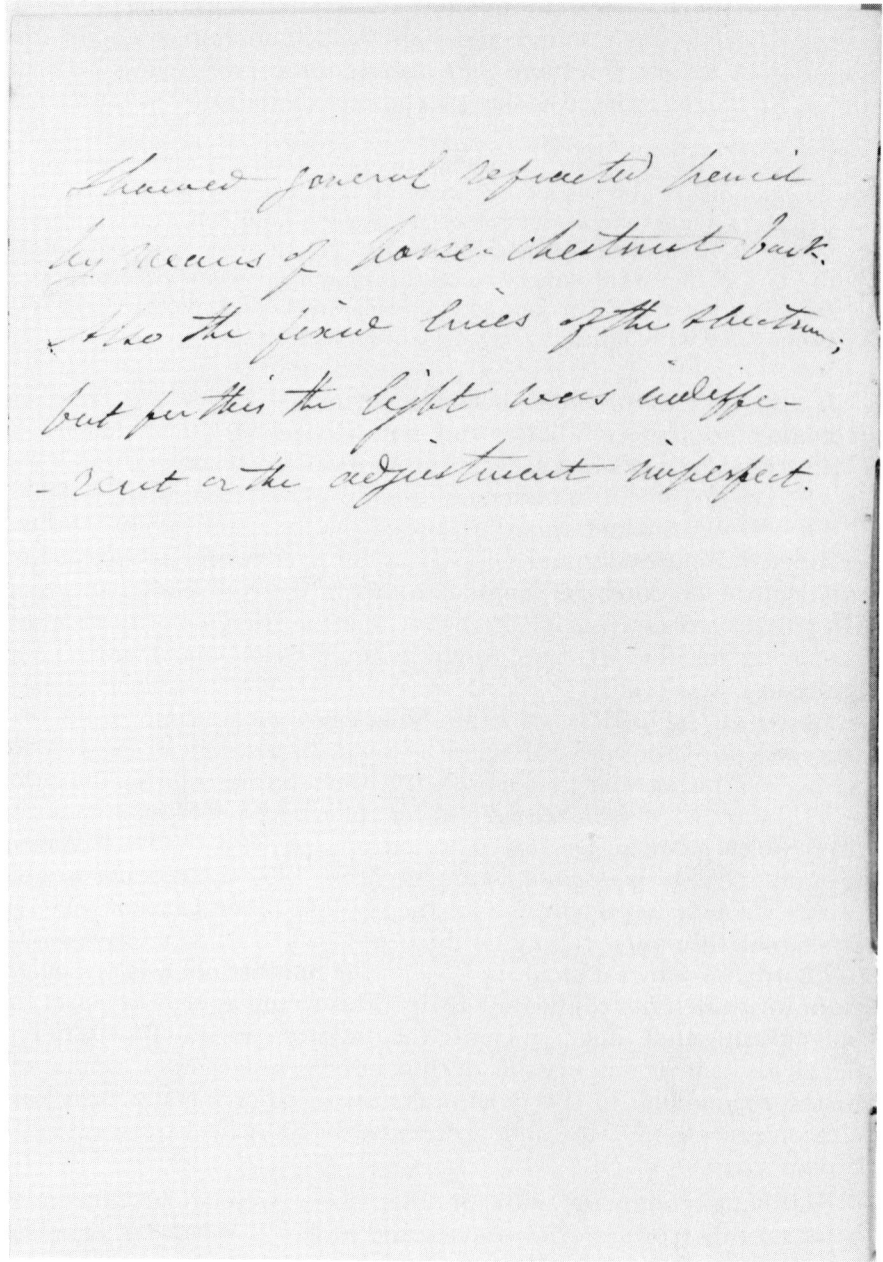

Figure 3.1 From Stokes's lecture notebook for 1852, 1857, and 1861 *CUL Stokes Collection* NB2. The notes are for his lecture of 18 April 1852. (By permission of the Syndics of Cambridge University Library.)

VI. Friday April 18 — Some item
Monday

1. Resume $\frac{dy}{dt} =$ ^{eg} of motion $\frac{1}{\rho}\frac{dy_x}{dt} = X - \frac{\partial^2 u}{\delta t^2}$

Transp $\frac{1}{\delta}$ $\frac{1}{\rho}\frac{dy}{dy} = Y - \frac{dM}{dt} - u\frac{du}{dx} - \cdots$

2. Necessity for the so called equation of
continuity.

3. First demonstration $\frac{d\rho}{dt} + \frac{d\cdot\rho u}{dx} + \cdots = 0$

4. Second demonstration $\frac{d\rho}{dt} + \rho(\frac{du}{dx} + \cdots) = 0$

5. Equations of condition relating to the
surface

6. Refer to Cauchy's dem? vide C.R. D. M. J.
vol 3
$=\{$ & owen to this $\}$

7. Interpretation : impulsive motion $\{W.H.\}$
Cork in water

8. Other interpr?: No ang? vel?

9. Apply gen? = ^{ns} to motion of rot. ?

10. Steady motion

11. Deduce it also from the = ^{ns} of
motion

man, Thomson had to sit facing Cayley, reading his work upside down.[16] Though the mathematician A R Forsyth, senior wrangler in 1881, valued his own attendance at Cayley's lectures, he too was the only undergraduate of his year present: 'Not for undergraduates was there to be attendance at Cayley's lectures, any one of which often contained results his research had obtained only since the preceding lecture; there he never gave a thought to the Tripos; and rarely indeed did the examiners pay the least heed to Cayley's work.'[17]

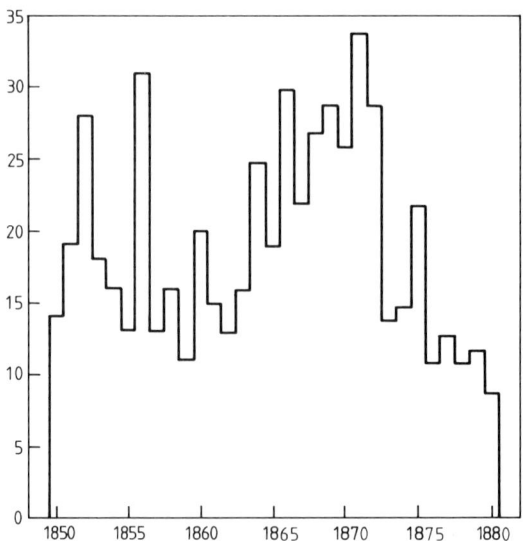

Figure 3.2 Enrolment in Stokes's class, 1850–80. From 1850–80 a total of 591 were enrolled (an average of 19 per year). For 1850–63 the total was 243 (average 17), for 1864–75 total 292 (average 24) and for 1876–80 the total was 56 (average 11). *Source:* Stokes's lecture-fee notebook *CUL Stokes Collection* NB1.

In his biography of J J Thomson, the fourth Lord Rayleigh—a student of Thomson's in the 1890s—thought J J Thomson was probably the only undergraduate of his year not only at Cayley's lectures but also those of John Couch Adams, who had replaced George Peacock as Lowndean professor of astronomy and geometry in 1859.[18] Karl Pearson, third wrangler in 1879, noted that Adams was 'used to small audiences'. Pearson deplored Maxwell's poor lecturing and did not finish the course.[19] J A Fleming, who had entered Cambridge at the age of 28 especially to work with Maxwell, treasured Maxwell's lectures but admitted: 'As a teacher Maxwell was difficult to follow. He had a paradoxical and allusive way of speaking, and he saw so much deeper into scientific problems than ordinary persons

that his utterances often appeared obscure.'[20] Fleming was one of two
students at Maxwell's lectures in 1878–9. J J Thomson, who entered
Cambridge in 1876, did not attend Maxwell's lectures, instead hearing
enthusiastic but unclear lectures on Maxwell's *Treatise* at Trinity
College given by W D Niven, third wrangler in 1866.[21]

By contrast, Forsyth found Stokes's lectures 'delightful', and Pearson
regarded him as one of the two best lecturers he ever heard.[22]
J J Thomson wrote:

> The lectures I enjoyed the most were those by Sir George Stokes on
> Light. For clearness of exposition, beauty and aptness of the experi-
> ments, I have never heard their equal. He had only the simplest
> apparatus at his command, no light but that of the sun, no assistant to
> help him. He prepared the experiments himself before the lecture and
> performed them himself in the lecture, and they always came off.[23]

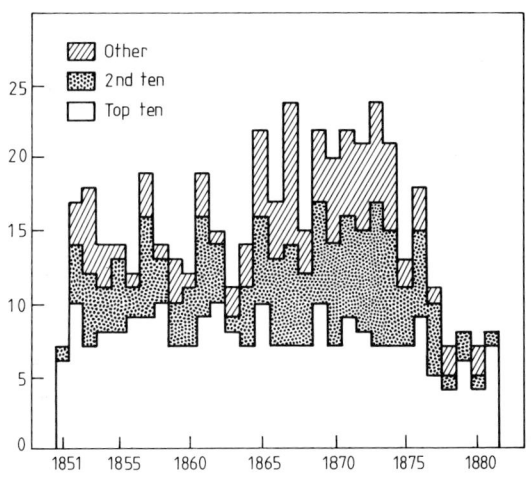

	1851–81	1851–64	1865–76	1877–81
Top ten	236 of 310 (76%)	115 of 140 (82%)	95 of 120 (79%)	26 of 50 (52%)
2nd ten	143 of 310 (46%)	53 of 140 (38%)	80 of 120 (67%)	10 of 50 (20%)
Other	100 of 547 (18%)	31 of 259 (12%)	64 of 231 (28%)	5 of 57 (9%)
Total	479 of 1167 (41%)	199 of 539 (37%)	239 of 471 (51%)	41 of 157 (26%)

Figure 3.3 Wranglers in Stokes's class, 1851–81. *Sources:* Stokes's
lecture-fee notebook *CUL Stokes Collection* NB1; J A Venn (ed)
1940–54 *Alumni Cantabrigienses* 6 vols part II 1752–1900
(Cambridge: Cambridge University Press); and J R Tanner (ed)
1917 *The Historical Register of the University of Cambridge* (Cambridge:
Cambridge University Press).

There remains the question of exactly why so many students attended Stokes's lectures. Like J M Wilson's tutor, both Pearson and Forsyth, for example, noted that the lectures did not pay for the tripos. However, hydrostatics, hydrodynamics, acoustics and optics *were* important MT subjects. Perhaps Pearson and Forsyth meant that the distinguishing feature of Stokes's course—the optical experiments—did not pay and that Stokes was not the only source for the other subjects, which indeed would have been covered more thoroughly by coaches than by Stokes. At any rate, it would appear that Stokes's own reputation, the quality of his lectures and their relevance to the MT combined to make his course a standard part of MT preparation for high wranglers for a quarter of a century.

Nevertheless, as figures 3.2 and 3.3 indicate, by the time Pearson, Forsyth and J J Thomson took the course, a decline had set in. Though the five years from 1877 to 1881 may seem too brief a period to be sure there was a decline, still the drop-off was abrupt and the number of high wranglers enrolled was appreciably lower than for any previous five-year period. Why? This was the time of the rise of the Cavendish and before the institution of the greatly reformed MT and NST in 1882. Maxwell's demonstrator began giving experimental lectures in 1876–7. Experimental lectures for the NST were available at Trinity College during the 1870s, and, as we have seen, J J Thomson attended lectures on electromagnetic theory at Trinity. Stokes's course was therefore no longer the only source of experimental lectures, and it did not cover the MT's new physical subjects of heat, electricity and magnetism. Evidently, some of those who would have taken his course were going elsewhere.

I have found only one list of Stokes's students for the 1880s, apparently for his class on physical optics in Lent Term of 1888.[24] The number of students was up to seventeen, but the composition of the class clearly shows the ways that curriculum changes were affecting Stokes's enrolment.

Most striking, only *two* of the seventeen were taking the class in preparation for the MT, one for the general part I, the other for the specialised part II. Obviously, Stokes's course was no longer an automatic part of preparation for the MT. Both W H Bragg (third wrangler in 1884) and A N Whitehead (fourth wrangler in 1883), for example, emphasised the narrowness of their MT preparation, neither mentioning Stokes's course.[25] Indeed, Bragg evidently read solely with Routh, not even attending J J Thomson's course at the Cavendish until after he had completed the MT.

By contrast, ten of the seventeen took Stokes's course to prepare for either part I or part II of the NST.[26] Part II of the NST required students to concentrate on one area but to show 'competent knowledge' of a second. Eight of Stokes's students from the 1888 class took part II, with four concentrating on physics and three electing it as their field of competent knowledge. Before 1880 only a sprinkling of NST students had taken Stokes's class, though Fleming took it in 1878 specifically to prepare for the NST of 1880.[27] By 1888 Fleming's use of Stokes's class, exceptional then, had become something of a norm. It was now tied much more closely to the NST than the MT.

In the absence of any of Stokes's class lists for the 1890s, one must be cautious. However, consider the following points. J J Thomson transformed the NST around 1890 into the principal Cambridge examination in physics. During the 1890s the Cavendish boasted a large staff and a considerable programme of classes in experimental and theoretical physics. The growing research at the Cavendish emphasised electricity. Electricity and magnetism were never important to Stokes's course, which continued to feature traditional optics, not Maxwell's electromagnetic theory of light. Indeed, when William Thomson urged Stokes to stand for the vacant Cavendish professorship in 1884, Stokes responded that he felt 'too old to take up a completely new set of subjects' and that 'those who would be under me would be knowing more about the subjects than I do myself'.[28] One can therefore surmise that, insofar as Stokes's course was relevant to the NST in the 1880s, it would have been more and more replaced by Cavendish courses as time went on. W C Whetham, who was in Stokes's class in 1888 and who received a first class in part II of the NST for 1889, did not mention Stokes in his memories of Cambridge days, instead noting the importance to him of the 'striking lectures' of J J Thomson and the 'advice and kindness' of R T Glazebrook, demonstrator at the Cavendish.[29] The chief physics graduate of the NST in the early 1890s was Nobel prize-winner C T R Wilson. In his reminiscences, he, like Whetham, discussed the Cavendish courses in physics, not Stokes's. He recalled Stokes himself mainly as a silent presence at meetings of the Cambridge Philosophical Society and the Cavendish Physical Society.[30] Moreover, insofar as Stokes's course was less germane to the MT in the 1880s than in the 1870s, it would have been even less useful in the 1890s, as Thomson shifted physics primarily to the NST leaving the MT for mathematicians and astronomers. Bertrand Russell (seventh wrangler in 1893), J H Jeans (second wrangler in 1898) and G H Hardy (fourth wrangler in 1898) all appear to have prepared for the MT mainly with their coaches and college lecturers, without the aid of Stokes's or any other professorial lectures.[31] Beneath Stokes's list of topics for his final lecture in Easter Term of 1893 there is even the poignant note: 'No one came'.[32]

Stokes's surviving records allow us to see how the changing Victorian context affected his students' choices of careers. In both the 1850s and the 1870s, for example, students' careers divided mainly among the church, secondary education, the law, fellowships in Cambridge colleges and university positions. The church and university positions saw the major shifts. Of Stokes's top students from the 1850s 27 per cent had at least a large part of their career in the church; of those from the 1870s, it was only 9 per cent.[33] On the other hand, whereas only 11 per cent of his top students from the 1850s secured university positions, 32 per cent from the 1870s did. Less than *one* per cent went into medicine or engineering.

Moreover, with Cambridge's proliferation of triposes, Stokes no doubt increasingly got students who were more directly interested in science. Almost all of his class of 1888, for example, went into education at the secondary or university level. None received a call from the bar or the church.

In the 32 years for which Stokes's class lists survive, there were at least 45 future fellows of the Royal Society in his class. His students included P G Tait (1851), Maxwell (1854), Routh (1854), R B Clifton, professor of physics at Oxford (1856), W K Clifford, professor of mathematics at University College, London (1865), James Stuart, professor of mechanism at Cambridge (1865), W H M Christie, Astronomer Royal (1867), G H Darwin, Plumian professor at Cambridge (1867), J Hopkinson, professor of electrical engineering at King's College, London (1869), H Lamb, professor of mathematics at Owens College, Manchester (1871), J Larmor, Stokes's successor as Lucasian professor (1879), and those mainstays at the Cavendish after Maxwell—Rayleigh (1864), R T Glazebrook (1875), W N Shaw (1875), J J Thomson (1879), L R Wilberforce (1888) and W C Whetham (1888).

In summary, we can see that curriculum changes, which Stokes generally favoured and sometimes specifically endorsed, eventually undermined his own prominence in Cambridge physics education. As the university and its physics education expanded, his own role diminished. He was never quite the centre of such studies, of course. That role belonged to Routh from the 1850s into the 1880s and to J J Thomson in the 1890s. Also, even in the early decades of his professorship, neither the institutional context nor his own personality engendered the sort of continuing teacher–student relationships enjoyed by Routh and J J Thomson.[34] It is nevertheless true, however, that for a third of a century Stokes's course was the Cambridge curriculum's most significant embodiment of that experimental–mathematical approach to physics which Cambridge's best physicists desired for the curriculum for so long. The mathematical side of Stokes's course made it relevant to the MT for three decades; the experimental, to the NST probably for one. The likes of Maxwell, Rayleigh and J J Thomson passed through Stokes's class on their way to leadership of mid- and late-Victorian physics. In short, with his support for reform and with his links to the MT and NST, Stokes was important to Cambridge physics education for a longer period of time than any other professor of the nineteenth century.

Thomson's Glasgow

Thomson's Glasgow changed much less than Stokes's Cambridge from 1846 to 1899. To be sure, there were changes, many of which were clustered in the 1870s. In 1870 the university moved into a new building, affording Thomson's natural philosophy class better facilities. An examination in natural philosophy had already been one of the requirements for the MA degree, and in 1872 a BSc degree was added with a similar examination as one of its requirements. In 1875 Thomson's nephew, James Thomson Bottomley, filled the newly created position of experimental demonstrator in natural philosophy, holding it until his uncle's retirement. As Thomson explained to Stokes in 1886, Bottomley gave many of the lectures and assisted Thomson in managing the laboratory of the natural philosophy class.[35] Though provisions in the Universities (Scotland) Act of 1858 had encouraged Scottish students to take degrees, it was not until the 1870s that the number of degrees awarded by Glasgow increased noticeably. Despite the various changes, however, physics education at Glasgow was always basically a one-man operation, in sharp contrast to Cambridge. It was Thomson's viewpoint which was imposed on the course and its content, as well as on his small staff.

In the early years of his professorship, Thomson combined various Glasgow pedagogical policies in shaping his natural philosophy course.[36] Like William Meikleham and John Anderson before him, he kept the mathematical and experimental parts of the class separate. Like Thomas Thomson, he added a laboratory where students in the class could carry out experimental work. Like his father, he introduced written examinations and offered various class prizes, in addition to the standard ones for 'general eminence', to encourage advanced work. The general eminence prizes were awarded by a vote of the class; the others by Thomson.[37] Advanced work in the natural philosophy class helped prepare a student for the examination for an MA degree with honours in the area of mathematics and natural philosophy. Problems on the honours MA examination required the calculus; those for an ordinary MA degree did not.

The undergraduate career of Thomson's best student in the 1850s, J D Everett, illustrates the system. In the 1855–6 session, Everett was in the senior mathematical class taught by James Thomson's successor, Hugh Blackburn, fifth wrangler in 1845. Everett was listed second among the students receiving prizes 'for general eminence in the business of the session' and first 'for excelling at examinations on paper during the session'. At the beginning of his final session, 1856–7, he won another mathematics prize in an examination open to members of the previous year's senior class. During the session he took Thomson's natural philosophy class and earned three more prizes. He was placed first in the list of students recognised for general eminence and won prizes for written examinations over the 'ordinary business of the class' and for his experiments 'on the conductivity of metals for electricity and on the electric convection of heat'. He won a university prize for the best essay on the topic: 'The Investigation of the Earth's Density'. At the end of the session he received an honours MA degree with 'highest distinction' in mathematics and natural philosophy. He was one of only 17 to receive a Glasgow MA in 1857 and one of only seven to receive one with distinction in mathematics and natural philosophy during the 1850s.[38]

Thomson's general course was a comprehensive treatment of the physical sciences which placed abstract dynamics and energy conservation at the centre. By 1870, he had developed a four-fold division for the class: '(1) abstract dynamics (including elements of physical astronomy), (2) properties of matter, (3) thermodynamics, and (4) illustration.'[39] The first three were a constant part of the course, but the subjects included under 'experimental illustration and demonstration' required more than one session to cover. At the outset of his professorship, Thomson had estimated that it would take more than two sessions to discuss all the subjects he wanted to.[40] In 1870, he stated in the *Calendar* that the subjects of illustration would fill two sessions, according to the following schedule:[41]

1870–71	1871–72
1. magnetism and electro-magnetism	6. capillary attraction
2. thermo-electricity	7. sound
3. elasticity	8. heat
4. hydraulics	9. electrostatics
5. light and radiant heat	10. galvanism and electrolysis

By 1873–4, the list was somewhat different (the kinetic theory of gases was included, for example), and the course was to extend over three years.[42]

A student's lecture notes from the 1862–3 session allow us to see how Thomson's scheme worked then. For the experimental part of the class, the notes record two series of lectures running through the session. The first included dynamics, properties of matter, energy, hydrodynamics and the dynamical theory of heat. It began in early November, encompassed 21 lectures by early December, and concluded with the 48th lecture in April. The second series, chiefly on electricity and magnetism, began in early December and, alternating with the first series, concluded with its 41st lecture in April. For the mathematical part of the class, Thomson delivered a third series of 60 lectures from November to April. Dynamics and thermodynamics were the principal topics, but Thomson also paid attention to hydro-dynamics and optics.[43]

Dynamics, Thomson declared in the 1862–3 session, was the science of force. Force was 'the name given to the cause of change of motion in bodies' and was directly sensed by us through 'muscular resistance'.[44] The primary qualities of matter relevant to abstract dynamics were 'extension, impenetrability, and inertia'. Gravitation, for example, was not an essential property. A man at the centre of the earth might never learn of gravity, though he would certainly recognise the first three properties. Abstract dynamics, then, was 'founded on force whatever its origin may be and resistance to joint occupation of space and inertia'.[45] The two divisions of dynamics were statics, where forces were in equilibrium, and kinetics, where forces produced motion.[46]

Natural philosophy's ultimate goal was the successful application of this science of force to the physical world, the reduction of physical phenomena to dynamics. Referring back to William Whewell's views of the 1830s, Thomson explained that astronomy was therefore 'a perfect science for practical purposes'. Magnetism, electricity, heat and chemistry, he explained, were in transition states. 'Hereafter they will be ranked under Dynamics. But we do not yet know enough about them to enable us to do this. We must first know what the electric matter is etc. All we know is that force is everywhere present.' Because of the successes of the undulatory theory of light and radiant heat, those areas were further advanced, but not so far as astronomy.[47]

Thomson regarded energy theory as a branch of kinetics. 'We shall view kinetic problems as illustrating energy.'[48] He explained that the 'action of a force setting a body into motion is generally a trans-formation of potential into kinetic energy'.[49] Hence, the enormous value of the doctrine of the conservation of energy was that it allowed one to deal with physical phenomena in the absence of exact knowledge of the underlying forces and matter involved. It would be especially useful for 'transition' sciences like electricity and magnetism, and, accordingly, Thomson's lectures on these subjects in 1862–3 often invoked energy considerations. Students that session heard about the relationships between electrical energy and heat, magnetic energy and work, chemical energy and heat, and electrical energy and work.[50] Of course, such energy relations also reflected Thomson's unified view of physical nature. As John Ferguson, a student in the 1859–60 session and later professor of chemistry at Glasgow, recalled:

> His impulse was to correlate phenomena and arrive at the principle underlying them, and this gave him a certain impatience with branches

of science which were still in the observational stage, and had not yet come under mechanical laws. Hence the most brilliant and weighty part of his course was at the end, when he summed up his teaching and generalised upon energy and the correlation of the physical forces, showed us Faraday's experiments on the conversion of electricity and magnetism, and Joule's conversion of work into heat.[51]

Thomson developed his optional experimental, or laboratory, class within a few years of assuming his professorship. He first listed prize-winners in the class for the session of 1854–5 and, for the next several years, also included the subjects on which the prize-winning work had been done. A partial list: thermo-electricity in 1854–5, electro-dynamics in 1855–6, electrical conductivity of alloys of copper and elasticity of copper in 1857–8, submarine telegraphic signals in 1858–9, electrostatic determination in absolute measure and atmos-pheric electricity and electrometers in 1859–60, contact electricity and the conducting power of solids for heat in 1861–2, submarine cables in 1864–5, electric resistances and voltaic batteries in 1865–6 and thermal conductivity of iron at different temperatures in 1866–7. The heavy emphasis was on electrical experiments with incursions into its thermal and chemical manifestations. The elasticity of metals was a continuing topic, and aspects of heat and magnetism appeared occasionally. Optics appeared not at all. In short, the list underscores the way Thomson assigned research problems devolving from his own physical and engineering researches.[52]

The form and content of advanced mathematical studies associated with the natural philosophy class varied from time to time. In the early years Thomson awarded prizes in the mathematical theory of electricity (1847–8), the mathematical theory of heat (1848–9) and the mathematical theory of electromagnetism (1849–50). These topics seem not to have been part of the formal class, but studied indepen-dently. In the early 1850s Thomson continued to offer prizes for performance in an examination on a specialised topic—in 1855–6: 'Images in physical science with special application to theories of heat and electricity'.[53] In the late 1850s and early 1860s, as his prize list for laboratory workers was growing, he awarded no prizes for mathema-tical studies. In the mid-1860s, however, he instituted a 'higher mathematical course' separate from the mathematical part of the regular course. The higher course evidently concentrated on a par-ticular topic each session—the theory of elasticity in 1865–6, the theory of vibrations and waves in 1866–7, waves of water in 1868–9 and vortex motion in 1885–6.[54] Adding the higher mathematical course was apparently Thomson's final solution to the problem of teaching mathematical physical theory (other than dynamics, which remained as the main mathematical part of the old course). A second-class honours MA degree required, among other things, an examination over 'the subject of the special course on mathematical physics for session of attendance'.[55] The role of independent study remained, as indicated by some of the books covered by the examinations for an MA degree, including for first-class honours Fourier's *Analytical Theory of Heat*. Thomson and Tait's *Treatise on Natural Philosophy* figured prominently in the course and in MA examinations from the time of its publication in 1867. Table 3.1 indicates changes in the content of MA examinations during the half century from the 1830s to the 1880s.

Consequently, throughout most of his professorship, Thomson provided opportunities for specialised mathematical and experimental work in addition to his main course with its own experimental and mathematical components. For the final third of the century, prizes for the higher mathematical course were listed along with the already established prize list for work in the laboratory. Also included, of course, were the 'general eminence' prizes, awarded by vote of the students. A student could combine an introductory overview of natural philosophy (composed of mathematical and demonstration lectures) both with further experimental work performed by himself and a higher mathematical study of some area of physics. But did they? Who were Thomson's students and what became of them.

Overall, nearly 7000 students enrolled in Thomson's class.[56] During the first 25 years of his tenure the class averaged about 85, generally increasing from around 80 at the start to just over one hundred. During his final 28 years, the class averaged around 175, reaching highs of around 230 in the 1880s and 1890s, before declining to under one hundred in Kelvin's final three years. Generally, nearly ten per cent of the students each year were included in the prize list for general eminence. The number of prize-winning students per year for experimental or higher mathematical work remained about the same after the early 1870s. Before that, as already mentioned, there were times when either no mathematical or no experimental prizes were given, and in the decade from 1861 to 1871 Thomson awarded a very large number of experimental prizes, possibly honouring nearly everyone who worked in the laboratory during that period. One notable trend was the increasing percentage of Thomson's prize-winning students who took degrees. That, incidentally, makes it easier to identify their careers, for the standard register of Glasgow's students in the second half of the nineteenth century is a roll of *graduates*.[57]

Let us look first at students listed for general eminence. Taking the first three for each year from 1847 to 1895 gives a total of 147 students. Of the 75 students from Thomson's first quarter century, 41 did not take degrees. Of those that did, 13 became clergymen and 10 found

Table 3.1 'Table of Subjects of Examination [in Natural Philosophy] for the degree of Master of Arts, According to the Different Classes of Distinction Which Candidates May Have in View.' *Sources: Glasgow University Calendar* for 1833–4, 1863–4 and 1881–2.

1833–4[a]

Minimum for Graduation	For Honourable Distinction	For Highest Distinction
Examination on the Subjects explained in the Class; Public Theme written in the Physic Class, with the Class Exercises given in every Monday, from the First of December during the Session, certified by the Professor, to be produced and examined. To those Candidates who may have been some time absent from College, the Professor will prescribe a set of Exercises.	In addition, all the other Exercises of the Classes, whether voluntary or not; with Examination on Newton's Principia, Book I. Sections I. II. III. and Book III. the whole.	Examination on the Mécanique Céleste of Laplace, and the whole of Newton's Principia; with perfect accuracy in all the branches of Examination, and proof of eminent talents and acquirements.

1863–4[b]

For Ordinary Degree	For Honours (Second Class)	For Honours (First Class)
Examination on the Subjects explained in the Class. On the Elements of Statics and Dynamics, with the solution of Problems not requiring the Differential or Integral Calculus; Experimental Science, including Heat, Electricity, and Magnetism; Herschel's Astronomy; Thomson and Tait's Elements of Natural Philosophy, First Part.	Certain portions of the Mécanique of Poisson, of Duhamel, or of Delaunay, the dynamical theory of heat; the solution of problems on all these subjects.	In addition, one or more of Airy's Tracts, on the Lunar Theory, the Planetary Theory, the Figure of the Earth; also, either Fourier's Théorie Analytique de la Chaleur, or the Mathematical theories of three of the following subjects, Electricity, Magnetism, Light, Sound.

1881–2

For Ordinary Degree

Examination on the Subjects explained in the Class. On the Elements of Statics and Dynamics, with the solution of problems not requiring the Differential or Integral Calculus; Experimental Science, including Sound, Light, Heat, Electricity, and Magnetism; Galbraith and Haughton's Astronomy; Thomson's Lecture on Navigation; Thomson and Tait's Outlines of Dynamics.

For Honours (Second Class)

Thomson and Tait's Elementary Treatise on Natural Philosophy; Maxwell or Balfour Stewart on Heat; Donkin on Sound; Central Forces.

The subject of special course on Mathematical Physics for session of attendance.

Portions of Tait's Thermo-dynamics, or of Thomson's Electrical Papers to be professed.

For Honours (First Class)

In addition, Portions of Thomson and Tait's Natural Philosophy (large work) to be professed; also, either Fourier's Théorie Analytique de la Chaleur, or the Mathematical theory of one of the following subjects, Electricity, Magnetism, Light, Sound, Elasticity of Solids.

For undulatory Theory of Light, Jamin, Cours de Phys., Vol. III, Part II, and Airy recommended.

---

[a] Meikleham's table of subjects reflects the emphasis he placed on short essays (exercises) by students on topics covered in the class. Some of these survive. See D B Wilson 1985 'The Educational Matrix: Physics Education at Early-Victorian Cambridge, Edinburgh, and Glasgow Universities' in P M Harman (ed) *Wranglers and Physicists: Cambridge Mathematical Physics in the Nineteenth Century* (Manchester: Manchester University Press) pp. 26–33.

[b] Thomson's table of subjects for 1863–4 came just after he began using a pre-publication version of elementary portions of his and Tait's *Treatise on Natural Philosophy* (published in 1872 as *Elements of Natural Philosophy*) and just before he introduced his special course on mathematical physics, which appeared in the table of subjects for a second-class degree for the first time in 1865–6.

university positions of one kind or another. Of the 72 from the next 24 years, only 11 did not take degrees. Fourteen went into secondary education and 13 found university positions. Only eight became clergymen, and there were four to six each in law, business, medicine, and engineering. The overall pattern, then, was for Thomson's top students increasingly to take degrees and enter a profession, with secondary education supplanting the ministry as their first choice.

During the same 49-year period, Thomson awarded mathematical prizes to 68 students. Of the 32 in the first period, 15 did not take degrees, and the others divided mainly among the church, secondary education, and universities. Only three of the 36 from the second period did not take degrees, and the largest numbers of the rest either obtained university positions (nine) or became engineers (six). Slightly fewer were in secondary education, medicine or the church.

Of the 163 students honoured for laboratory work in the first period, 94 did not take degrees, 40 became clergymen, and 14 went into medicine. Of the 65 students from the second period, 33 did not take degrees, 12 became engineers, and only four went into the church.

There was little overlap between the mathematicians and the laboratory workers. Only 11 students in these 49 years won prizes in both areas. In the first 25-year period, the mathematicians tended to go into the church or education, the experimentalists into the church or medicine. Among graduates of the second period, more of the mathematicians went into engineering than before, but the mathematicians were more evenly divided among professions than were the laboratory workers. The greatest change from the first period to the second was the increased number of laboratory workers going into engineering. Indeed, these figures help demonstrate the well known contributions of Thomson's laboratory to electrical technology, especially submarine telegraphy. As one of his students, who substituted for him in 1869–70, wrote:

> The success of the Atlantic Cable is in great measure the result of years of patient work in the Glasgow Laboratory. . . . Students are trained in Glasgow for home and foreign telegraphic service, and at the laying of the French Atlantic Cable two of the best practical and scientific electricians were young men selected from the Glasgow class.[58]

Undoubtedly, many of the laboratory workers without degrees were going into the expanding telegraphic industry. 'There is quite an epidemic amongst the laboratory students of desire to become *telegraph engineers*', wrote Thomson to a correspondent.[59]

Table 3.2 lists 62 of Thomson's top students who gained some prominence in physics, mathematics, chemistry or engineering.

Table 3.2 Kelvin's students prominent in physics, engineering, mathematics, or chemistry.[a]

Student[b]	Gen. Em.[c]	Math.	Expt.	T.E.S.[d]	Glasgow degrees[e]	FRSE	FRS	Career[f]
James Porter	46–7				MAII47			9th wr. 51; Master, Peterhouse
W J Steele	46–7 47–8				MAII47			2nd wr. 52; d. 1855
John Kerr	46–7 47–8 48–9				MAI49		1890	Math. lect., Free Church Training Col., Glasg.
Donald McFarlane	47–8							Thomson's assistant
C A Smith	52–3				MAI53			2nd wr. 58; Controller, Auditor-General of the Cape
William Jack	52–3				MAI53	1875		4th wr. 59; Prof. math., Glasg. U.
Frank Maclean	54–5	54–5					1895	Engineer
J D Everett	56–7		56–7		MAI57	1863	1879	Prof. nat. phil., Queen's C, Belfast
John Aitken[g]	56–7				MAI57	1875	1889	R. Napier & Sons, shipbuilders; Private laboratory, Falkirk
John Ferguson			59–60		MAI62	1888		Prof. chem., Glasg. U.
R K Miller	60–1				MA62			1st Smith's pr. 67; Prof. appl. math., Royal Naval Col.
J Y Buchanan	62–3					1870	1887	Chemist, physicist on *Challenger* expedition; Lect. geography, Camb. Un.
W F King	66–7 67–8		64–5 66–7			1880		Chief eng., Western & Brazilian Teleg. Co.; King & Co, Leith
Peter Alexander[h]	65–6	65–6			MA70			Private tutor in Glasg. in math., physics, & mechanical and engineering science
William Ross	66–7				MA67			Prof. math., Free Church Christian Col., Madras

(continued)

Table 3.2 (*continued*)

Student[b]	Gen.[c]	Em.[c]	Math.	Expt.	T.E.S.[d]	Glasgow degrees[e]	FRSE	FRS	Career[f]
W E Ayrton	67–8		67–8	67–8				1881	Prof. phys., elec. eng., Central Technical Col., London
J D H Dickson	67–8 68–9		67–8	67–8		MAII70	1876		5th wr. 74; Thomson's assistant; Fellow, tutor, Peterhouse
Thomas Muir	67–8						1874	1900	Math. & sci. master, Glasg. H.S.; Superintendent-General of Educ. Cape of Good Hope
Dugald Mackichan[i]	68–9		68–9	68–9	70–2	MAII70	1926		Vice-Chancellor, Bombay University
Henry Dyer	68–9			71–2	72–3	MA73 BSc 73 DSc90			Prof. engineering, Tokyo
T Lindsay Galloway	69–70		72–3	72–3	73–4	MAI73	1918		Mining engineer, Kintyre
M M Pattison Muir	70–1			70–1			1873		Fellow, head of chem. lab., Caius Col.
William Ramsay	70–1							1888	PhD Tübingen; Prof. chem., Univ. Col., London; Nobel prize 1904
William Bottomley	71–2			71–2					Thomson's nephew and assistant
A W Thomson	72–3 73–4					BSc74 DSc91			Prof. eng., Royal Agricultural Col.
J J Dobbie[j]	74–5					MA75	1904	1904	Prof. chem., Univ. Col. of North Wales
Andrew Gray	74–5					MAII76	1883	1896	Prof. nat. phil., Glasg. U.
J C Watt	74–5					MAI75			10th wr. 80; Fellow, Jesus Col.
James Thomson	74–5 75–6			74–5		MAII79			Civil eng., naval architect; Lect., Durham Col. of Science; Thomson's nephew
Thomas Gray	74–5			74–5	76–9	BSc78	1880		Thomson's assist.; Prof. eng., Rose Poly. Inst.
Arthur Smithells[k]	77–8							1901	Prof. chem., Leeds U.

Name				Degrees			Position
Archibald Barr	75–6			BSc78 DSc90 LLD13	1921	1923	Prof. civil eng., mechanics, Glasg. U.
J M Dodds	76–7			MAI78 BSc79			4th wr. 81; Fellow, tutor, Peterhouse
R F Muirhead	76–7	77–8		MAI81			Lect. math., Mason Col.
Alexander Russell	78–9	81–2		MAI81	1906	1924	14th wr. 85; Principal, Faraday House
G A Gibson	79–80	82–3		MA81 LLD05	1890		Prof. math., Glasg. and W. of Scot. Technical Col.; Prof. math., Glasg. U.
John M'Cowan	81–2	81–2	84–5	BSc83 MAI86 DSc92			Demonst. physics, assist. math., Univ. Col., Dundee
Magnus Maclean	83–4	83–4	84–5	MAI86 DSc95	1888		Prof. elec. eng., Royal Technical Col., Glasgow
L Crawford	84–5	85–6		BSc86 DSc99	1903		5th wr. 90; Prof. math., U. of Cape Town
John Dougall	84–5	85–6		MAI86 DSc	1921		Lect. math., Royal Technical Col., Glasg.; Blackie & Sons, publishers
A Galt[1]				BSc87	1888		Keeper, Technological Dept., Royal Scottish Museum
Abraham Levine	87–8	88–9		MA89			12th wr. 92; Actuary
J R Erskine-Murray	87–8	90–1		BSc92 DSc97	1897		Marconi & Co.; Clark, Ford & Taylor, submarine cable engineers
J D Cormack	87–8	89–90		BSc92			Prof. civil eng., mechanics, Glasg. U.
James Holm	88–9			MAI91			Prof. physics, South Africa U.; d. 1897
J H T Tudsbery	88–9			BSc89 DSc95			Sec., Institution of Civil Engineers
J B Henderson	88–9		91–2	BSc92			Prof. appl. mech., Royal Naval Col.

(continued)

Table 3.2 (*continued*)

Student[b]	Gen. Em.[c]	Math.	Expt.	T.E.S.[d]	Glasgow degrees[e]	FRSE	FRS	Career[f]
D R Boyd	88–9				BSc92			PhD Heidelberg; Prof. chem., Univ. Col., Southampton
Peter Pinkerton	89–90				MAI90 DSc09	1905		Head math. master, George Watson's Col., Edinb.; Rector, Glasg. H.S.
H S Carslaw	89–90				MAI91			4th wr. 94; Prof. math., Sydney U.
R M Ferrier	90–1				BSc91			Lecturer, Durham Col. of Science
D S Jerdan	91–2				MA92	1900		PhD Heidelberg; Business
J S Dunlop	92–3 93–4				MAII96			31st wr. 99; worked in Glasg. U. phys. lab.; d. 1901
J W Peck	92–3	93–4			MAI95	1904		Demonstrator expt. phys., Glasg. U.; Secretary, Scottish Educ. Dept.
James Muir	93–4	96–7			BSc96 DSc02			Camb. BA (research) 99; Prof. nat. phil., Royal Technical Col., Glasg.
G D Valentine	94–5	95–6			MAI96			9th wr. 99; Sheriff-substitute, Perthshire
M M'C Fairgrieve			94–5	95–6	MA	1910		Camb. nat. sci. tripos 1st class 99; Senior science master, Edinb. Acad.
John Miller		95–6			MAI96 DSc09	1910		Prof. math., Royal Technical Col., Glasg.
R J T Bell	96–7	97–8		96–7 98–00	MAI98 DSc11	1916		Prof. math., U. of Otago, New Zealand
R M Dyer	97–8	98–9			BSc00			Naval architect
D K Picken	97–8	98–9			MAI99			6th wr. 02; Master, Ormond Col., U. of Melbourne
F P H Stirling[m]	98–9				MA01			12th wr. 03; Prof. math., Madras Christian Col.; d. 1906

a *Sources*: W I Addison (ed) 1898 *A Roll of the Graduates of the University of Glasgow, from 31st December 1727 to 31st December 1897* (Glasgow: James MacLehose & Sons); *Glasgow University Prize and Degree Lists, 1834–1863*; *Glasgow University Calendar* (Glasgow, 1864–1900); *University of Glasgow Class Catalogue* (Glasgow, 1874, 1876, 1885); J A Venn (ed) 1940–54 *Alumni Cantabrigienses* 6 vols part II: 1752–1900 (Cambridge: Cambridge University Press); *The Record of the Royal Society of London* 4th edn (London: The Royal Society, 1940); *Proceedings, Transactions*, and *Year Book of the Royal Society of Edinburgh*; *Dictionary of National Biography*; *Who Was Who*. I am grateful for assistance especially from Professor J M A Lenihan and also from Dr William Duncan, Executive Secretary of The Royal Society of Edinburgh and Mrs Alma Topen of the Glasgow University Archives.

b Thomson's Japanese students are not included. With the year in which they appeared in the prize list for general eminence given in parentheses, they were: N Tamiguchi (76–7), R Masuda (76–7), R Shida (80–1), N Takayama (80–1), K Watanabe (84–5), B Mano (86–7) and E Odagiri (97–8). Shida was a Thomson Experimental Scholar for 1881–2. They appear to have become engineers rather than physicists. The three for whom Addison gives information (Tamiguchi, Masuda and Watanabe) practised engineering in Tokyo. None appears in the table of 21 'principal members of the first generation of Japanese physicists' in K Koizumi 1975 'The Emergence of Japan's First Physicists: 1868–1900' *Hist. Stud. Phys. Sci.* **6** 105. However, 2 of the 21 (M Goto and A Tanakadate) did study physics at the University of Glasgow, presumably in the 1890s, but their names are not in the prize lists for the natural philosophy course. At least three of Thomson's students taught at the Imperial College of Engineering in Tokyo: H Dyer, engineering, 1873–83; W E Ayrton, natural philosophy and telegraphy, 1873–8 and T Gray, telegraphy, 1878–81. Thus, as with his class as a whole, Thomson's Japanese connection was apparently directed as much or more towards engineering than physics.

c The first three columns give the sessions for which students appeared in Thomson's prize lists.

d In 1869 Thomson contributed £1000 to establish three Thomson Experimental Scholarships worth £20 per year. There were sometimes four per year during the 1880s.

e Throughout the period students could try for an ordinary MA degree or one with first- or second-class honours, though the requirements varied (see table 3.1). These degrees are denoted by MA, MAI, and MAII.

f The table does not try to list every position a man held. I have included Cambridge degrees, not only to indicate what Cambridge connection there was for Thomson's students but also as a certain level of prominence in itself.

g Aitken won a prize as a 'private' student for the 1856–7 session. As distinct from 'public' students, private students were not subject to examinations or exercises and, therefore, could not count the class towards requirements for a degree.

h Addison lists Alexander as a professor of physics at Anderson's College, but the *Calendars* for Anderson's College do not list him as having been on the faculty.

i While holding his Thomson Experimental Scholarship, Mackichan did research on the ratio of electrostatic to electromagnetic units which was published in *Phil. Trans.* (1873) 409–27.

j Dobbie was in Thomson's class list for 1873–4 but did not appear in the prize lists. He received BSc and DSc degrees from Edinburgh in 1878 and 1879, respectively.

k Smithells was in Thomson's class list for 1875–6 but did not appear in the prize lists.

l Galt was in Thomson's class list for 1884–5 but did not appear in the prize lists.

m Stirling also received a prize in the 1899–1900 session for a special course in geometrical optics taught by G D Valentine.

Presumably, though Thomson was the only professor of natural philosophy at Glasgow, he was not the principal influence on the careers of some of them, for example, the chemists, Ferguson, Ramsay and Smithells. The table clearly indicates that Thomson's class was a substantial source of manpower for the growing professions of science and engineering. Yet, it is also clear that, with the possible exception of W J Steele, who died at 23, none of Thomson's students even remotely approached his own power as a natural philosopher. Thomson's own rejection of Maxwell's electromagnetic theory must have contributed to the dearth of his students from 1875 on who even reached the level of fellow of the Royal Society of London. Thomson's students were about as likely to gain prominence in mathematics, chemistry or engineering as in physics. There were only seven fellows of the Royal Society whose work touched on physics. John Kerr is the most famous of the seven, having the 'Kerr effect' named after him. Kerr was a lecturer in mathematics at the Glasgow Free Church Training College for Teachers and published on two important experimental discoveries in the 1870s. One showed a connection between light and electricity, the other a connection between light and electromagnetism (the Kerr effect). The Royal Society of London awarded him a Royal medal in 1892. J D Everett was professor of natural philosophy at Queen's College, Belfast, from 1867 to 1897. Though he published some experimental researches on sound and light, he was chiefly an author of textbooks. His translation and heavy revision of Deschanel's *Natural Philosophy* gained wide usage as a textbook in experimental physics.[60] Thomson used it in his class for many years. John Aitken, initially planning to be an engineer, decided because of poor health to pursue physics instead. The research he performed in his private laboratory on the role of dust particles in the formation of fog won him a Royal medal in 1917. Andrew Gray, Kelvin's successor, served as his assistant for a decade and then as professor of physics at University College, North Wales, from 1884 to 1899. Like Everett, he was best known for his textbooks. J Y Buchanan left Glasgow intending to follow chemistry and was appointed chemist and physicist for the *Challenger* expedition of the 1870s. His resultant chemical and physical studies of sea water helped establish the science of oceanography. Alexander Russell published mainly on alternating currents and for 30 years was principal of Faraday House, the college of electrical engineering in London. Sent by the Indian telegraphic service to study with Thomson, W E Ayrton was, like Russell, primarily an electrical engineer. His electrical researches earned him a Royal medal in 1901. With him as professor of physics and electrical engineering, the Central Technical College of London became a centre of electrical engineering education.

In summary, there were various kinds of students enrolled in one part or another of Thomson's natural philosophy class. There were several thousand generally educated students. There were some who did well in Thomson's class but became prominent in other fields— mathematics, chemistry, medicine. Many went from Thomson's laboratory into engineering. In physics, there were only a few who actually did combine mathematical and experimental studies at a high level to take full advantage of that sophisticated combination of experimental and mathematical physics that constituted Thomson's natural philosophy class. A few students joined the small cluster of assistants around Thomson. Some won wider acclaim as physicists, including three Royal medals altogether. However, none was a physicist, either experimental or mathematical, of the first rank.

Glasgow and Cambridge

Both Thomson and Stokes lectured predominantly to students who would not become physicists. Even most of their best students went into other careers, their choices of careers being influenced by surrounding cultural changes. There were several contrasts between their classes, however. In most respects, the contrasts reflected continuing differences between Glasgow and Cambridge Universities.

In the first place Stokes, unlike Thomson, was always competing with other physics teachers of great ability. Hence, the place of Stokes's course in the Cambridge curriculum was more affected by developments within his university than was Thomson's. As a consequence, Stokes's course was never the principal locus of physical instruction, and the importance it originally possessed declined with the rise of the Cavendish.

Second, their mix of mathematics and experimentation differed. With a class largely of high wranglers to be, Stokes could present an integrated experimental–mathematical course. With a class mainly of non-mathematicians, Thomson followed Glasgow tradition and kept the two separate. 'He very rarely introduces an experiment into his mathematical course, or mathematics into his experimental course.'[61]

Third, the content of their classes differed. Shaped by the early-Victorian MT, Stokes and his class concentrated on fluids, sound and light. Rooted in the *milieu* of early-Victorian Glasgow University, Thomson's more comprehensive course embraced these subjects plus dynamics, heat, electricity and magnetism. Both courses necessarily contained much of the elementary and routine, but they also both had their deeper theoretical slants. Here, Stokes presented an ether physics grounded in the undulatory theory of light; Thomson an energy physics which encompassed the whole of physical nature and suggested future reduction to a dynamical theory of force and matter. Neither Stokes nor Thomson made much of Maxwellian electromagnetic theory. J J Thomson did, of course, thereby contributing to the Cavendish's growing prominence late in the century.

Fourth, in contrast to Cambridge and Stokes, there was the practical quality of Glasgow and its university and of Thomson's career and those of his students. Cambridge produced relatively few engineering graduates and did not create an engineering tripos until the 1890s, despite the presence of interested professors and the addition of other triposes earlier. In Glasgow an engineering chair was established in 1840, and it was probably the engineering side of Thomson's career that best harmonised with the aptitude and inclinations of his natural philosophy students. Table 3.2 has somewhat more the character of engineering than physics, and the engineers—like Ayrton and Russell—seemed to represent the sort of engineering science embodied in Thomson's own work. Of course Thomson was not alone in this, as he was joined on the faculty by engineering professors of like mind: Lewis Gordon (1840–55), W J M Rankine (1855–72), and his own brother, James (1873–89).[62] The many mutual reinforcements here are obvious: an industrial city, a practical university, professors who combined science and engineering and students who followed engineering careers at various levels of prominence. Indeed, this cluster of reinforcing factors may even have profoundly shifted the centre of gravity of Scottish physics education from the natural philosophy–moral philosophy dominance of Edinburgh University to the natural philosophy–engineering dominance of Glasgow during William Thomson's tenure as professor. In any case, it was a Glasgow tradition that undoubtedly nurtured Thomson's engineering interests and, in turn, was strengthened by his brilliant achievements.

Fifth, there is the different impact of their respective students on the science of physics. For three or four decades, at least until the Cavendish established itself, Thomson presented the most modern physics education in Britain. Yet he produced no first-rate physicists. For three decades Stokes presented a well attended course popular with high wranglers. Though more limited than Thomson's course, passing through it, nevertheless, were an abundance of future fellows of the Royal Society, including much of the leadership of mid- and late-Victorian physics. With its strong practical tradition in medicine and engineering, Glasgow evidently did not attract the best pro-spective natural philosophers, despite Thomson's presence. The Englishman, J D Everett, chose to attend Glasgow, but his country-men J J Thomson and Rayleigh did not. Nor did the Irishman, Larmor, nor even the Scots, Tait and Maxwell. They did not choose Cambridge because of Stokes's presence, however. Rather, it was the famous Cambridge tripos that attracted them. And the attraction was augmented by the scores of tripos graduates throughout English secondary and higher education urging bright students to go in for the MT. It was, for example, a former senior wrangler at Owens College who advised J J Thomson to go to Cambridge.[63] Of course, William Thomson was himself a former wrangler sending students to Cambridge. However, none of the 17 men in table 3.2 who took Cambridge degrees was a leading physicist. Evidently, the pattern of a Scottish undergraduate's becoming a Cambridge wrangler and then attaining the highest distinction in physics began and ended in the days of Thomson and Maxwell. There was, at any rate, a great contrast between the accomplishments of Thomson's and Stokes's respective *physics* students, but it cannot be attributed to a difference in Thomson's and Stokes's own abilities in teaching and research. Rather, as with other contrasts, it mainly represented a difference in universities, as institutional reputation overrode professorial prestige.

Furthermore, the careers of Thomson's students help underscore the singularity of his own genius. Being able to trace Glasgow's influences on Thomson's developing thought obviously does not mean that the Glasgow setting necessarily turned students into great physicists. In fact, Thomson was the only Glasgow undergraduate of the entire century to become a major physicist. But he also studied at Cambridge. Ironically, even though his classes at Glasgow probably influenced his physical researches more than did his preparation for the MT, his name still contributed to the MT's high reputation which drew later students there, and away from Glasgow.

Finally, looking ahead to subsequent chapters, we should note the historiographical importance of Stokes's and Thomson's lectures. These chapters frequently probe for views not extensively discussed by Stokes and Thomson in their publications. Hence, the significance of their lectures, along with other unpublished material, as evidence for what they thought about things. As one small example, speaking of the 'dreadful ravages' that would confront humans in a world of instantly melting snow and ice, Thomson proclaimed to his class in 1847: 'In the necessity of latent heat we see the goodness and kindness of God.'[64]

---

1   Stokes was also professor of physics at the Government School of Mines during the 1850s, but this chapter deals with his Cambridge position.

2   A Russell 1938 *Lord Kelvin* (London and Glasgow: Blackie and Son) p. 38. Other accounts of Thomson by former students include: J D Cormack 1899 'Lord Kelvin: A Biographical Sketch' *Cassier's Magazine* **16** 3–17, 133–57; W Jack 'The Late Lord Kelvin' (undated newspaper clipping in Glasgow University Archives); A Gray 1908 *Lord Kelvin: An Account of His Scientific Life and Work* (London: J M Dent) pp. 279–98; W E Ayrton 1908 'Kelvin in the Sixties' *Popular Science Monthly* **72** 259–68; J Ferguson 1908 'Lord Kelvin: A Recollection and an Impression' *Glasgow University Magazine* **20** 276–82; M Maclean 1908 'Lord Kelvin' *Proc. R. Phil. Soc. Glasgow* **39** 60–72; D Wilson 1910 *William Thomson, Lord Kelvin: His Way of Teaching Natural Philosophy* (Glasgow: John Smith and Son); D Murray 1927 *Memories of the Old College of Glasgow: Some Chapters in the History of the University* (Glasgow: Jackson, Wylie) pp. 119–41; and [A W H Hedderwick] 'Glasgow University in the 'Seventies: Random Recollections of an Old Student' reprinted from *Glasgow Herald* (January–March 1933) pp. 21–6 in *GUL*.

3   See G Kitson Clark 1967 *An Expanding Society: Britain 1830–1900* (Cambridge: Cambridge University Press). There is a rapidly expanding literature on the Victorian scientific community. For an overview see D L S Cardwell 1972 *The Organisation of Science in England* 2nd edn (London: Heinemann).

4   See D A Winstanley 1947 *Later Victorian Cambridge* (Cambridge: Cambridge University Press); L Stone 1974 'The Size and Composition of the Oxford Student Body 1580–1910' in L Stone (ed) *The University in Society* 2 vols (Princeton: Princeton University Press) II, 6, 91; Murray *Memories of the Old College* p. 282 (see note 2); *The Curious Diversity. Glasgow University on Gilmorehill: The First Hundred Years* (Glasgow: University of Glasgow Press, 1970) p. 54; J D Mackie 1954 *The University of Glasgow, 1451–1951. A Short History* (Glasgow: Jackson and Co.) pp. 213, 277, 287, 305; and W I Addison (ed) 1898 *A Roll of the Graduates of the University of Glasgow: From 31st December, 1727, to 31st December, 1897* (Glasgow: James Maclehose and Sons) pp. 680–1.

5   For a discussion of the developing curriculum at Cambridge in physics see D B Wilson 1982 'Experimentalists among the Mathematicians: Physics in the Cambridge Natural Sciences Tripos, 1851–1900' *Hist. Stud. Phys. Sci.* **12** 325–71.

6   Stokes to Maxwell, 18 February 1871, *Maxwell Papers CUL* Add. MS 7655 II 41.

7   G D Liveing 'Appreciation [of Stokes]' *Memoir* I 95.

8   Stokes, Autobiographical essay *Memoir* I 8.

9   Stokes to W Thomson, 23 October 1849 *CUL Kelvin Collection* S349.

10   Stokes, Lecture notebook for 1852, 1857, 1861, *CUL Stokes Collection* NB2.

11   *Ibid.*

12   Stokes, Lecture notes for 1900 *CUL Stokes Collection* PA276.

13   Lord Rayleigh, Notes on Stokes's lectures for 1864 American Institute of Physics Niels Bohr Laboratory, Microfilm of Rayleigh Papers, box II reel 3.

14 Stokes, Lecture-fee notebook 1850 to 1880 *CUL Stokes Collection* NB1.

15 G Howson 1982 *A History of Mathematics Education in England* (Cambridge: Cambridge University Press) p. 126.

16 J J Thomson 1936 *Recollections and Reflections* (London: G. Bell and Sons) p. 47.

17 A R Forsyth 1935 'Old Tripos Days at Cambridge' *Mathematical Gazette* **19** 163.

18 Lord Rayleigh 1942 *The Life of Sir J. J. Thomson* (Cambridge: Cambridge University Press) p. 9.

19 K Pearson 1936 'Old Tripos Days at Cambridge, As Seen from Another Viewpoint' *Mathematical Gazette* **20** 31–2.

20 J A Fleming 1934 *Memories of a Scientific Life* (London and Edinburgh: Marshall, Morgan, and Scott) p. 61.

21 Thomson *Recollections* p. 42.

22 Forsyth 'Old Tripos Days' p. 162 (note 17); Pearson 'Old Tripos Days' p. 32 (note 19).

23 Thomson *Recollections* p. 48.

24 Stokes, Lecture notebook for 1888, 1890, and 1893 *CUL Stokes Collection* NB20. The list of students' names is with notes for lectures in January and February 1888.

25 W H Bragg, Autobiographical notes, in G M Caroe 1978 *William Henry Bragg, 1862–1942: Man and Scientist* (Cambridge: Cambridge University Press) pp. 22–4; A N Whitehead 1951 'Autobiographical notes' in P A Schilpp (ed) *The Philosophy of Alfred North Whitehead* 2nd edn (LaSalle, Illinois: Open Court) p. 7.

26 Of the ten five took parts I and II of the NST and attended Stokes's lectures before taking part I, one took Stokes's class after part I of the NST but before part II, two took it after part I of the MT and before part I of the NST, and two took it after part I of the MT and before part II of the NST. Another three students had already finished their examinations before taking Stokes's course. One student took no tripos, and I have been unable to identify another. Stokes's students thus illustrate the variety of ways of studying physics at Cambridge after 1882.

27 J A Fleming *Memories* p. 68.

28 Stokes to W Thomson, 6 December 1884 *CUL Kelvin Collection* S466.

29 W C Dampier (formerly Whetham) 1950 *Cambridge and Elsewhere* (London: John Murray) pp. 17, 51.

30 C T R Wilson 1960 'Reminiscences of My Early Years' *Not. Rec. R. Soc.* **14** 163–73.

31 B Russell 1956 *Portraits from Memory and Other Essays* (New York: Simon and Schuster) pp. 62–5; B Russell 1967 *The Autobiography of Bertrand Russell* vol. I 1872–1914 (New York: Simon and Schuster) p. 90; E A Milne 1952 *Sir James Jeans: A Biography* (Cambridge: Cambridge University Press) p. 4; and G H Hardy 1926 'The Case Against the Mathematical Tripos' *Mathematical Gazette* **13** 64–5.

32 Stokes, Lecture notebook for 1888, 1890, and 1893 *CUL Stokes Collection* NB20.

33 'Top students' means those who were among the first ten wranglers.

34   For example Rayleigh apparently felt that Stokes did not fully respond
       to his eagerness as a student. (R J Strutt, Fourth Baron Rayleigh 1924
       John William Strutt, Third Baron Rayleigh (London: Edward Arnold)
       pp. 37–8.)
35   Thomson to Stokes, 8 November 1886 *CUL Stokes Collection* K278.
36   For a discussion of the early years of Thomson's class, see D B Wilson
       1985 'The Educational Matrix: Physics Education at Early-Victorian
       Cambridge, Edinburgh and Glasgow Universities' in P M Harman
       (ed) *Wranglers and Physicists: Cambridge Mathematical Physics in the Nine-
       teenth Century* (Manchester: Manchester University Press) pp. 12–48.
37   Thomson explained his practice, including his use of *viva voce* and
       written examinations, in a letter to William Walton, who was gathering
       such information for consideration by the Board of Mathematical
       Studies at Cambridge. During the six-month session, Thomson used a
       'mixed' system of written and oral examinations, on the basis of which
       the class would vote for their outstanding peers. Thomson thought one
       could determine merit in a brief period, however, only with a written
       examination. (Thomson to Walton (1859) Loose sheets in Minute Book
       of the Board of Mathematical Studies, Cambridge University Archives,
       Min. V. 7. This is a copy of Thomson's letter in Walton's hand.)
38   *Glasgow University Prize and Degree Lists, 1834–1863; Glasgow University
       Calendar* (1901) p. 358; and Addison *Roll of Graduates* (note 4) p. 680.
39   *Glasgow University Calendar* (1871) pp. 36–7.
40   Thomson to J D Forbes, 22 November 1846 *St Andrews University Library
       Forbes Papers.*
41   *Glasgow University Calendar* (1871) pp. 36–7.
42   *Glasgow University Calendar* (1874) p. 39.
43   David Murray, Notes on Thomson's three series of lectures for 1862–3
       GUL MSS Murray 325, 326. Murray discussed Thomson's class in his
       history of the university, reporting that he wrote out his notes for the
       class in consultation with a classmate, John Y Buchanan, later a fellow
       of the Royal Society of London. (Murray *Memories of the Old College of
       Glasgow* (note 2) pp. 119–31.)
44   Murray, Notes on Thomson's lectures, 10 and 11 November 1862 *MS
       Murray* 325.
45   *Ibid.* 13 November 1862.
46   *Ibid.* 4 November 1862.
47   *Ibid.* 6 November 1862. Murray recorded Thomson's reference: 'See
       Whewell address BAR (1832), XIII'. The date is off by one year; on
       page xiii of his address to the British Association in 1833, Whewell
       stated: 'Astronomy . . . is not only the queen of the sciences, but, in a
       stricter sense of the term, the only perfect science.' (Whewell 1833,
       Address *Br. Assoc. Rep.* p. xiii.)
48   Murray, Notes on Thomson's lectures, 17 November 1862 *MS Murray*
       325.
49   *Ibid.* 27 November 1862.
50   Murray, Notes on Thomson's lectures, 19 December 1862, 14 and
       28 January 1863, 6 and 18 February 1863, 3 March 1863, 22 April 1863
       MS Murray 326.

51   Ferguson 'Lord Kelvin' (note 2) p. 281.

52   *Glasgow University Prize and Degree Lists; Glasgow University Calendar.* In his Bakerian Lecture 'On the Electro-Dynamical Qualities of Metals' in 1856, for example, he acknowledged assistance from his assistant, Donald McFarlane, and four students (*MPP* II 189). In 1885 Thomson discussed the early years of his laboratory, reporting that about 25 students per year worked in it and that about three-fourths of them became ministers. ('The Bangor Laboratories' *PLA* II 483–9.)

53   *Glasgow University Prize and Degree Lists.*

54   *Glasgow University Calendar; Thompson* II 851.

55   *Glasgow University Calendar* (1869) pp. 64–5.

56   Lists of students enrolled in Thomson's class for each year are in *Catalogus Togatorum in Academia Glasguensi* and *University of Glasgow Class Catalogues.*

57   Addison *Roll of Graduates* (note 4).

58   R K Miller 1870 'The Proposed Chair of Natural Philosophy' *Cambridge University Reporter* (23 November) 118.

59   Thomson to Jessie Crum, 29 March 1872, in *Thompson* II 622. Also cited by R Sviedrys 1976 'The Rise of Physics Laboratories in Britain' *Hist. Stud. Phys. Sci.* **7** 414. Sviedrys discusses the importance of Thomson's students' laboratory for Britain's science-based telegraphic industry.

60   A Privat Deschanel 1877 *Elementary Treatise on Natural Philosophy* translated and edited with extensive additions by J D Everett 4th edn (London: Blackie and Son).

61   Miller 'The Proposed Chair of Natural Philosophy' (note 58) p. 119.

62   See L Gordon 1847 *A Synopsis of Lectures to Be Delivered Session 1847–48* (Glasgow: Griffin); T Constable 1877 *Memoir of Lewis D. B. Gordon, F.R.S.E.* (Edinburgh, private circulation); W J M Rankine 1856 *Introductory Lecture on the Harmony of Theory and Practice in Mechanics* (London and Glasgow: Richard Griffin); W J M Rankine 1857 *Introductory Lecture on the Science of the Engineer* (London and Glasgow: Richard Griffin); J B Henderson 1932 *Macquorn Rankine, Professor of Civil Engineering and Mechanics in the University of Glasgow, 1855 to 1872* (Glasgow: Jackson, Wylie); and D F Channell 1982 'The Harmony of Theory and Practice: The Engineering Science of W J M Rankine', *Technology and Culture* **23** 39–52.

63   Thomson *Recollections* p. 30.

64   Anonymous, Notes on William Thomson's lectures, 1847–48 *GUL* MS Gen. 633.

4

Religion and Science

Within the familiar phrase 'science and religion' lurks the danger of deep misunderstanding. It might, for example, be taken to mean that there have usually existed two clearly distinguishable activities or areas of knowledge, with one well defined group of people pursuing one, another the other. It might call to mind images of the trials of Galileo and John Scopes, thereby seeming to confirm an inevitably adversarial relationship between the two areas of knowledge and their respective adherents. In historical reality, however, there have not been two absolutely demarcated areas of knowledge or groups of people, and conflict is only one of the relationships that has existed. Throughout history, people have found innumerable ways of combining a wide variety of ideas concerning such subjects as nature, God, Jesus and the Bible.

Even focusing on a particular time and place can produce a complex picture. In nineteenth-century Britain, for example, clergymen were some of the leading interpreters of nature, and professional scientists were among those speaking out on religious issues. There was a spectrum of religious thought that changed from decade to decade. Scientific thought was hardly monolithic either, and it too changed during the century. Most important to recognise is the absence of a sharp dichotomy between religion and science. Study of nature could prove God's existence which, in turn, could influence thought about both the Bible and the basic structure of the physical world. Understanding the development of life on earth could depend on both reading the Bible and examining fossils.

Long gone is the time when one could regard Victorian scientists simply as non-religious opponents of religion. Gone also is the time when one could be content merely with pointing out that many scientists were Christians. There now exist several studies showing that scientists' ideas reflected the diversity of religious thought in the culture around them. Especially helpful in this regard, and in providing an introduction to the ideas of Thomson and Stokes, is a paper by Michael Ruse, 'The relationship between science and religion in Britain, 1830–1870'.[1] Discussing several major scientists of that period, Ruse describes a range of views from conservative to liberal. All were Christians who believed that nature displayed God's design and operated in accordance with God's laws. There were great differences in emphasis, however. Conservatives thought the Bible contained some degree of objective information about the early history of man, emphasised design in nature as proof of God's existence, and affirmed God's proclivity to intervene miraculously in natural phenomena. Liberals thought Scripture provided no help in understanding man's early history, emphasised the existence of natural laws as proof of God's existence, and thought God worked through law, not miracle. With such views defined well before publication of Darwin's *The Origin of Species* in 1859, conservatives tended to reject even the occurrence of evolution while liberals tended to accept not only the reality of evolution but also that it had occurred in accordance with Darwin's theory of natural selection. For one group, Darwin contradicted God's personal involvement in nature; for the other, he provided a splendid example of God working his will through natural processes. A typical 'centrist' response to Darwin coupled a willingness to accept evolution with the view that natural selection insufficiently allowed for God's design in nature and that God had probably intervened in the origin of man.

A generation younger than the men discussed by Ruse, Stokes and Thomson reached maturity in the 1830s and 1840s when those in Ruse's study, like John F W Herschel and William Whewell, were prominent spokesmen. Though Stokes and Thomson agreed with one another in many ways, they also disagreed, along the general lines described by Ruse. They both thought that the Bible was God's revelation, that William Paley's design argument was valid, that God had created life, that God had designed man's mind to be suitable for acquiring scientific knowledge, that not all of Darwin's ideas were correct, and that what Stokes called 'directionism' offset any materialistic views allegedly supported by modern science. Neither discussed the results of German Biblical scholarship or, so far as I know, responded directly to the two major works in English conveying that liberal approach to the Bible as a work to be understood in its own historical context—*Essays and Reviews* (1860) and *Lux Mundi* (1889). They differed in their emphases on Scripture and in their specific responses to Darwin, with Stokes generally following a conservative line, Thomson more of a centrist. Such labels should be qualified, of course, in light of the dramatic shift in religious views of British intellectuals during the century. By 1900, Thomson's 'centrist' position looked rather conservative; Stokes's conservatism exceedingly conservative.

This chapter and the next explore Stokes's and Thomson's ideas in these areas. The next chapter on methodology includes their views of God's design of the human mind. This chapter contains four sections. The first examines the religious character of the respective contexts within which Stokes and Thomson matured. The next deals with Stokes's support of the doctrine of 'conditional immortality', the most significant product of his intense search of Scripture for religious truth. The third discusses Stokes's and Thomson's respective responses to Darwin's theory of evolution. The concluding section considers their 'directionist' alternative to materialism and their conviction that physics was superior to biology and geology, to the benefit of theology.

The religious background

Though men are not simply the products of their backgrounds, Thomson and Stokes did retain the general theological perspectives of the 1830s and 1840s into the early twentieth century. Moreover, their differences paralleled differences in their respective backgrounds.

For the more conservative Stokes, an upbringing in an Anglican Evangelical home had a lasting effect. As the vicar of the church in Cambridge where Stokes had long been church warden declared: 'Though he was never narrow in his faith and religious sympathies, he always held fast by the simple evangelical truths he learned from his father, the Protestant rector of Skreen, in the county of Sligo.'[2] Indeed, the general evangelical movement was the vital element in early-nineteenth-century British religion. Crossing denominational lines, it swelled membership in Anglican and dissenting churches. Unlike high-church or liberal Anglicanism, it was noted for neither sophisticated theological argument nor a historical and scholarly approach to the Bible. Rather, evangelicals emphasised emotional attachment to Christianity, earnestness of religious purpose, close attention to Scripture, and deep concern for the doctrine of Atonement with the attendant threat of eternal punishment for the unsaved. To save those so threatened, they enthusiastically evangelised for their version of Christianity, founding, for example, the Church Missionary Society in 1799 and the British and Foreign Bible Society in 1804. As an adult, Stokes was active in both of these societies and wrote on doctrinal matters relevant to missionary work.[3]

The strength of Anglican Evangelicalism at Cambridge may even have helped persuade Stokes to enter there. Prevailing religious views at Cambridge, however, had more to do with Archdeacon William Paley (1743–1805) than with Evangelicalism. All three of his chief works (*Natural Theology, Evidences of Christianity* and *Moral Philosophy*) received wide attention at Cambridge, the latter two being required for Cambridge examinations.

Paley's mutually reinforcing *Natural Theology* and *Evidences* rejected David Hume's arguments against the design argument and the reality of miracles. Paley discussed numerous examples of design in a watch-like universe and confidently proclaimed the existence of a one, good, designing God. Furthermore, 'once believe that there is a God, and miracles are not incredible'.[4] Miracles constituted the primary evidence for the truth of Christianity. Where Hume saw unreliable testimony for the reality of miracles, Paley spoke of the 'probity and good sense' of the witnesses to Christ's miracles and of the hardships they endured as they sought to spread His message.[5] Their willingness to suffer for what they preached helped demonstrate the truth of Christianity. 'Auxiliary evidences for Christianity' included Biblical prophecy, the high morality of the gospels, the candor of the writers of the New Testament, the originality of Christ's character, and the agreement of the four gospels on the nature of Christ's character. As one well known Cambridge graduate of the 1820s summarised his (and doubtlessly many others') reaction to Paley's books: 'I am convinced that I could have written out the whole of the *Evidences* with perfect correctness, but not of course in the clear language of Paley. The logic of this book and as I may add of his *Natural Theology* gave me as much delight as did Euclid.'[6]

Paley's *Moral Philosophy*, however, though a required part of the examination for a pass degree, did not give everyone quite the same 'delight' that his other books had Darwin. Paley presented a utilitarian morality which during the 1830s was challenged by the Cambridge professors Adam Sedgwick and William Whewell, both conservatives discussed by Ruse. They asserted that morality rested on a Divinely implanted innate sense of right and wrong. As a new undergraduate in November 1837, Stokes attended the four university sermons in which Whewell attacked Paley's moral philosophy.[7]

Stokes's religious thought combined these several elements, with Scriptural truth providing the essential ingredient. Though he disagreed with the usual evangelical view of the doctrine of eternal punishment, for example, he did so in typical evangelical fashion, through close investigation of Scripture. In moral philosophy, Stokes followed Whewell, rejecting utilitarian morality, which had only a 'very limited application', in favour of the 'school of moralists [who] hold that we have an innate consciousness of right and wrong'.[8] Moreover, he regarded the design argument 'much in the same way that was mentioned long ago by Paley in his *Natural Theology*'[9] and, like

Paley, declared: 'Admit the existence of a God, of a personal God, and the possibility of miracles follows at once.'[10] This conclusion strengthened the 'evidences' for Christianity provided by the Bible's 'honestly-written narratives' of Jesus's life and character.[11] Stokes approached the Bible with both an *Evidences* concern for the 'head' and an evangelical concern for the 'heart'. 'The evidence for the resurrection of Jesus Christ is never to be separated from a consideration of the character and teaching and works of Jesus Christ. The head and the heart must go together'.[12]

Thomson's background, in addition to Cambridge, included his family and his teachers at Glasgow University. His father had initially intended to become a Presbyterian minister, and according to Thomson's sister, 'Our father read with us regularly every Sunday morning some chapters in the Old Testament, and in the evening some in the New.'[13] Thomson's brother, James, eventually abandoned Presbyterianism to become a Unitarian and decided to leave Glasgow in 1849, writing to a friend that 'when my father died, I felt that I could not live longer with the other members of our family withholding from them my real sentiments'.[14] James gave up revealed religion for more of a natural religion. 'I can view what we call the Laws of Nature', he wrote to his friend, 'in no other light than merely as expressions of the will of an Omnipresent and Ever Acting Creator.'[15] Thomson's professors of logic and of moral philosophy at Glasgow, Robert Buchanan and William Fleming, were both Presbyterian ministers. Buchanan's lectures on logic included a section on the evidences of Christianity, and Fleming argued that God had created the happiest of all possible worlds, with each member of it being designed with faculties suitable for its place in the world in order to produce as much happiness as possible. Thomson's notes on Fleming's lectures include passages like: 'Goodness means benevolence or a spirit which delights to promote happiness. Look to [the] nature of man. God has made us what we are—sentient beings. Our very existence is a proof of his kindness.'[16] The year Thomson took the natural philosophy course at Glasgow, William Meikleham declared that not only did the study of natural philosophy extend our power over nature and gratify the mind, 'above all, it leads us to view the Creator as the Great First Cause, and as maintaining the energies of nature'.[17] Not long into the course, the ailing Meikleham gave the lectures over to a man greatly influential on the young Thomson, Glasgow's professor of astronomy, John Pringle Nichol. In a series of books on astronomy, Nichol argued, for example, that the genius of someone like Kepler, unable to be satisfied during an earthly lifetime, foreshadowed a 'necessary Immortality'.[18] Study of the intricate workings of the solar system, Nichol thought, allow us to 'exclaim, as we humbly contemplate, THE FINGER OF GOD IS THERE'.[19]

The views fashioned by Thomson within this context, less evangelical than Stokes's, included both revealed and natural religion, but with much less emphasis on the Bible than Stokes placed. However, surviving evidence suggests not only that Thomson was knowledgeable about the Bible, but that his views did have a distinct Scriptural component. When his brother, James, concluded that God had provided 'no other revelation of Himself but that contained in His wonderful works, and in the ideas which He has implanted in our minds, or rather which He has adapted our minds to receive', Thomson 'reasoned very strongly with his brother in favour of the evidence for revelation'.[20] In a draft of a paper on thermodynamics in 1851, Thomson linked the dissipation of energy in the universe with the Biblical pronouncement that 'the earth shall wax old as doth a garment'. He included, in the draft, a long footnote on miracles, writing, for instance, 'The vision of Peter and some of the circumstances connected with it are, I think, satisfactory illustration of such miracles.'[21] Like Stokes, Thomson endorsed the message of Paley's *Natural Theology*, speaking in 1871 of the 'solid and irrefragable argument so well put forward in that excellent old book'.[22] His niece summarised: 'The Bible he reverenced; he had studied it with care and was at home in it, but he seldom spoke of religious matters, or entered into arguments about them.'[23]

In conclusion, the evangelically influenced Stokes was both more conservative than Thomson and more 'evangelical' in actively addressing theological issues. Indeed, Stokes was probably the most public religious scientist of the late-Victorian period, frequently speaking before Church Congresses and the Victoria Institute, of which he was president. One discovers Thomson's views, on the other hand, in occasional published passages and in scattered remarks in his correspondence and unpublished manuscripts. Whereas Stokes agreed to give 20 Gifford Lectures on natural theology in the 1890s and—though struggling—managed to do so, Thomson declined the same invitation, overmodestly declaring, 'All that I could say on the subject can be said in five minutes and I have already said it.'[24] By amply displaying Stokes's outspoken conservatism, the case of the doctrine of conditional immortality, the sort of topic ignored by Thomson, perfectly illustrates the contrast between the two.

Conditional immortality

Stokes remembered 'when I was a little child being so horrified at the idea of endless torments that I wished there was no God and no future state, lest I should fall into them.'[25] The young George Stokes was not the only one unhappy with the doctrine of eternal punishment. At times it must have seemed likely to become a source of eternal controversy. In 1853 F D Maurice reinterpreted the doctrine so that 'eternal' did not mean a period of time. He was dismissed from his professorship at King's College, London. In the 1860s great and official Church consternation greeted criticism of the doctrine in *Essays and Reviews* and by Bishop Colenso of Natal in connection with his missionary work. The agnostic Charles Darwin, in his *Autobiography* written around 1880, called it 'a damnable doctrine' and could 'hardly see how anyone ought to wish Christianity to be true'. His Christian wife's response was no less hard on the doctrine itself: 'Nothing can be said too severe upon the doctrine of everlasting punishment for disbelief—but very few now [1882] wd. call that "Christianity", (tho' the words are there).'[26] The urban masses, their tenuous link to organised religion already documented by the religious census of 1851, began hearing the militant atheism preached by the working-class Secular or Freethought Movement which was formed in the 1850s and peaked in the 1880s. 'When Secularists named the Christian doctrines they found most objectionable, they were almost always those of Eternal Punishment, Hell, the Atonement, and Damnation for unbelievers.'[27]

Against this background, the movement for the doctrine of conditional immortality as a replacement for that of endless punishment began in the 1840s and was much discussed during the 1870s. Most widespread among Congregationalists and Evangelical Anglicans, the view reserved immortality for the saved, condemning the rest to extinction. Though less cruel than eternal torment, extinction retained the sting absent from the 'universalist' alternative that everyone would eventually be saved. The doctrine was seen as befitting a kind God who nevertheless expected people to behave themselves. At first met only with reprobation, conditionalists succeeded well enough to contribute significantly to the well known overall changes in Victorian religious thought. At the end of the century, 'it is true that the doctrine of hell had not been removed from the official theological confession of any denomination, but men were no longer deprived of office for teaching a tentative universalism or regarded with suspicion for espousing the doctrine of conditional immortality'.[28]

Stokes began his mature deliberations on eternal punishment in 1851 when he was 32 years old. He was possibly influenced by the book *Life in Christ* published in 1846 by Edward White, a Congregational minister who became the leading proponent of the doctrine of conditional immortality. By the late 1860s Stokes had fully accepted the doctrine. In the mid-1870s he began what became an enormous correspondence with Edward White himself, and White quoted at length from one of Stokes's letters in the preface of his book's third edition in 1878. In the early 1880s Stokes was caught up in a long correspondence on behalf of Walter Dening, a missionary to Japan for the Church Missionary Society who evidently had been preaching conditional immortality too vigorously. In connection with that controversy Stokes published a pamphlet privately in 1882 which presented three missionaries' favourable reports on the doctrine that had been forwarded to Stokes by White.[29] He wrote to the Church Missionary Society that 'if this counsel or this work be of God, ye cannot overthrow it, lest haply ye be found even to fight against God. Ye cannot indeed overthrow it; but how know you but that in the endeavour to stifle it by muzzling the evangelist sent out to the heathen you may be the occasion of preventing the salvation of many a soul for whom Christ died?'[30]

Finally, in the 1890s, Stokes published his own formulations of the doctrine, endorsing several points defined by Edward White and others whose works had flourished in the 1870s. They thought, first, that truth in these matters depended ultimately on the Bible. 'This book rests the question of Immortality wholly on interpretation of Scripture', White wrote.[31] 'One verse of the Bible on the nature of man, on the source of his life, on the meaning of his death, must outweigh a whole treatise of Plato, Aristotle, or Epicurus', another declared.[32] Second, the doctrine of conditional immortality was verified by Scripture whereas neither universalism nor the doctrine of eternal punishment was. Third, the tripartite nature of man—that he consisted of body, soul and spirit—was Biblical; the dual nature of man was Platonic, and wrong. Fourth, related questions—for example, whether men were conscious, alive, and/or evangelised during the intermediate state between death and Judgement—were worth examining, even at length, but in the end were far less crucial than the question of immortality. Hence, theirs was an intricate and exhaustive search of Scripture for truth—a search driven by challenges to Christianity at home and abroad and from all levels of society. The same search, with its conceptual and social motivations, carried over into Stokes's writings.

As the true counter to the false doctrine of eternal torment, the concept of conditional immortality was Stokes's principal concern. That man consisted of something besides ponderable matter—something which could therefore be immortal—Stokes thought was shown by ordinary experience. We retain our personal identities, as evidenced by memory, even though the matter of our bodies continually changes. At the end of a fainting spell, even a long one, our thoughts can flow continuously from their state before the spell, even though our always changing material bodies must have altered to some extent during the interval of unconsciousness. The corollary of materialism, determinism, is so contrary to our sense of free will that the whole point of view flies 'completely in the face of common sense'. Materialism was possible, Stokes thought, only for someone who thought the activity of ponderable matter sufficient to explain everything. But such well established scientific entities as luminiferous ether, magnetic force and gravitation demonstrated the insufficiency of ponderable matter and mechanical principles by themselves. Such 'mysterious' agents seemed to require 'something which we must regard as superadded to the mechanical properties of matter'.[33] However, though 'the materialistic hypothesis' was inadequate, so also was what Stokes called 'the psychic theory', according to which man consisted of body and soul, soul being inherently immortal and hampered in its functions because of its union with body. This view also, Stokes argued, foundered on ordinary experience. When the body became incapacitated—through, for example, fainting, injury or near drowning—the soul was not liberated to think more clearly, quite the contrary. Moreover, there existed no good, non-Biblical arguments in favour of the psychic theory.[34]

More important for Stokes, neither was there Biblical authority for innate immortality. The message of Biblical passages dealing with eternal life was that it was for the redeemed, the rest meeting extinction. In John 3:16, for example, Jesus stated: 'For God so loved the world, that He gave His only begotten Son, that whosoever believeth in Him should not perish, but have everlasting life.' Thus, unbelievers perished; they did not suffer the infinite misery of eternal punishment. The concept of conditional immortality, the view of *Life in Christ*, provided a basic insight into the Bible's true meaning.

> It is wonderful what harmony it introduces, both between one part of Scripture and another, and between the teaching of Scripture and what commends itself to our moral sense. As a friend of mine in Cambridge, before whom I brought that view, said to me, 'Reading the Bible with that idea in the head is like turning a key in an oiled lock'.[35]

The theory of man's tripartite nature, more ambiguous than the notion of conditional immortality, attempted to explain man's present make up and to understand what part of man survived death. Again, there were two parts to the discussion, Biblical and non-Biblical.

The non-Biblical argument focused on the already mentioned insufficiencies of the psychic theory and involved the limitations of soul evidenced by its close association with thought and consciousness which were so frequently absent. In his Gifford Lectures, Stokes, combining parts of the psychic and materialistic theories, postulated a deeper theory incorporating a third component:

> May it not be there is a something constituting the *ego* which, on the one hand, is not to be identified with thought, and which may exist

while thought is in abeyance; while, on the other, it is not to be identi-fied with ponderable matter, but yet exercises over ponderable matter a sort of command? May it not be that thinking is a process which results from the interaction of the ego on the organism with which the ego is associated, over which it is, as it were, placed in command? According to this view, the ego is something lying deeper down in our nature than thought itself—something the destruction of which is not involved in the destruction of the body, inasmuch as it does not consist of ponder-able matter—something which might conceivably, without any breach of continuity, preserve the personal identity between the man who died and the same man in some different stage of existence.[36]

Stokes speculated that thought would survive as a function of ego. Ego might be able either to think by itself or through interaction with the 'future body' which 'is promised according to the Christian religion'.[37]

The argument to this point may appear somewhat thin. Has not Stokes needlessly conflated soul and thought, thus artificially creating the requirement for a third entity? Could not soul rather easily be made to serve the functions that Stokes reserved for ego? Regarding soul's connection to body, does not the cessation of thought in an incapacitated body support Plato's view more than Stokes's? Could not a Platonist readily answer Stokes's objections? However such questions might be answered, the historical point to be realised is that any weakness in the above arguments would, for Stokes, simply demonstrate the limitations of natural theology. The real source of knowledge in these matters was not reason, but revelation.

Precluded from discussing Scripture in the Gifford Lectures, Stokes stated merely that, contrary to Plato's theory that man consists of body and soul, 'in Scripture we have a threefold division, into body, soul, and spirit'.[38] The exegetical problem was to distinguish between spirit and soul. On the back of a letter to him, Stokes listed nearly two dozen verses from the New Testament under the heading 'spirit $\neq$ soul'.[39] In print, he cited the creation of man in Genesis where 'we meet with the expression that God breathed into man's nostrils, after he was formed, "the breath of life, and man became a living soul"'. Though one should not be naively literal about this passage, never-theless, this 'breath' or 'spirit' according to Stokes,

> is spoken of as a sort of energy, the interaction of which with the material organism produced a living being. It is represented therefore, not so much as a living thing, but rather that which lay at the very basis of life, something deeper down even than very thought itself.[40]

Moreover, Stokes pointed out, in Scripture 'when that in man which is not put an end to by death is spoken of, it is not, I think called "soul", but "spirit"'.[41] For example, Luke 23:46: 'And when Jesus had cried with a loud voice, he said, Father into thy hands I commend my spirit: and having said thus, he gave up the ghost.'[42]

Hence, Stokes thought that man's personal identity resided in his spirit, or ego, not his soul. Spirit produced manifestations of life, like thought, through interaction with the body–soul component of the organism. Spirit survived death, and those spirits meriting eternal life were provided, presumably at the time of the Judgement, with another body. 'What the nature of this body may be we do not know, but we are pretty distinctly informed that it will be something very different from that of our present bodies. . . .'[43] Though evidence regarding man's state between death and Judgement was 'exceedingly meagre', Stokes leaned to the view that during that time life persisted but thought did not. Stokes was undoubtedly attracted by the fairness of this view, for each person, no matter how long before the Judgement he died, would be unaware of the passage of time and thus relieved of a long or short period of anxious waiting. We would perceive ourselves to 'be brought immediately face to face with our final account to receive our final destiny'.[44] Again, fainting afforded an analogy: 'I told you I knew from my own experience how very curiously time appears to be annihilated so long as one is in a faint'.[45]

The doctrines of man's tripartite nature and of his unconscious state between death and Judgement were not nearly so important as the doctrine of conditional immortality. It was this last that dealt with the baneful doctrine of eternal punishment, Stokes's childhood bogeyman which now was alienating from Christianity members of his own social–intellectual *milieu* as well as members of the working class. Yet, the doctrine of man's tripartite nature did more than help clarify questions of immortality, for in doing so it struck another blow at Platonism's pernicious infiltration into Christianity.

Evolution

Stokes and Thomson responded to the theory of evolution from their respective religious viewpoints. They agreed that science ought to explain as much as possible, but differed on how much had in fact been explained scientifically in the instance of evolution. Frequently referring to the Bible, Stokes questioned whether evolution had even occurred, especially in the case of mankind. Thomson said altogether much less than Stokes about evolution. With no invocation of Scripture, Thomson, like John Herschel, was evidently willing to accept that evolution had occurred but thought Darwin's concept of natural selection ignored God's design in nature. He did not discuss the issue of human evolution. They did agree that God must have created life, though even here it was Thomson who suggested a possible naturalistic explanation of the origin of life on earth. We can see their differences and similarities by considering the issues comprising the overall question of evolution: the evolution of the solar system; the history of earth and its life; and the origin of man.

Like most Victorian scientists, both Thomson and Stokes thought that the evolution of the solar system was correctly explained by Laplace's nebular hypothesis, according to which a nebula of non-luminous matter slowly condensed, first becoming luminous matter and eventually contracting sufficiently to form the sun. The earth and the other planets formed from bits of the hot, luminous matter left behind in the process of condensation. Hence, while the large, luminous nebula slowly shrank into a sun, the small, molten earth was more quickly cooling towards its present condition.

It was Stokes who tried to reconcile all this with Scripture. He focused on the Genesis story that light appeared on the first day of creation well before the sun which appeared on the fourth. To begin with, as Stokes wrote in 1864, though 'it is not the office of the Bible to teach us natural science', nevertheless, because 'here the writer professes to describe what could not be cognizable by man', Genesis must be true 'in some sense'. The 'days' did not have to be 24-hour days, for example, especially since Genesis spoke of days prior to the sun's existence.[46] Accordingly, Genesis could be reconciled with the nebular hypothesis in at least two ways. First, 'it seems probable that the earth would have made considerable progress in its cooling, and what depends upon it, before the luminous matter inside its orbit would have collected into a definite sun'.[47] The events of the first three days could thus have occurred while the luminous matter was still coalescing to form the sun on the fourth. Or, it was possible that the sun did form from a nebula on the first day but that initially it was not distinctly visible from the earth, only 'appearing' on the fourth day when the earth's steamy atmosphere had cleared.[48] Either way, the scientific account agreed with the Biblical narrative. Light appeared on the first day and was able to nourish vegetation appearing on the third, with the sun coming on the fourth. That other details might not fit the scheme so well bothered Stokes little: 'But if we suppose that the record in Genesis was meant for the people of the time, . . . then it would be preposterous to demand scientific accuracy of detail. A general rough accordance is all that we ought to expect; and that I think we have'.[49] The sense in which Scripture was true thus lay somewhere between 'slavish literalism' and scientific accuracy.

The nebular theory also indicated a long history for the earth from molten mass to life-filled globe. Geology demonstrated how much the earth's surface and its life had changed. How, then, was one to understand the origin and development of life?

Thomson, writing five years before the publication of Darwin's *Origin*, evidently reflected the widespread view that God had intervened on successive occasions to create life, various plants and animals, and mankind:

> Purely mechanical reasoning shows a time when the earth must have been tenantless; and teaches us that our own bodies, as well as all living plants and animals, and all fossil organic remains, are organized forms of matter to which science can point no antecedent except the Will of a Creator, a truth amply confirmed by the evidence of geological history.[50]

By the late 1860s, however, Thomson's earlier emphasis on discontinuity in the geological record of life had been replaced by talk of continuity and evolution. In his presidential address to the British Association in 1871, he even appeared to suggest a possible continuity between life elsewhere in the universe and the origin of life on earth. He stated that meteoric fragments from the ruins of other life-filled worlds may have brought simple forms of life to earth and 'according to the doctrine of continuity . . . all creatures now living on earth have proceeded by orderly evolution from some such origin'.[51] What Thomson did object to was Darwin's theory of natural selection, both because his research on the age of the earth showed that the earth had not been habitable long enough for such a slow process to have been effective, and because natural selection did 'not sufficiently take into account a continually guiding and controlling intelligence'.[52] Thomson was 'profoundly convinced that the argument of design has been greatly too much lost sight of in recent zoological speculations'.[53]

One should not overestimate Thomson's naturalism or his support for evolution, however. In the 1871 address, for example, he said he felt that the hypothesis of natural selection 'does not contain the true theory of evolution, *if evolution there has been, in biology*'.[54] Moreover, despite Thomson's apparently favourable presentation of his theory of the meteoric origin of earth's life, Stokes reported that Thomson's purpose 'has been a good deal misunderstood', for he had not so much been supporting his own theory as opposing an even more naturalistic one:

> He was not attempting to actually account for the origin of life on our earth, but *merely to show to biologists* what the nature of a real cause might be, in contradistinction to mere conjectures as to the spontaneous origin of life, which is all that those who hold to such an origin have to show for it.[55]

Indeed, Thomson may have been responding specifically to Huxley's statement in his address the year before that, if he were able to look upon the earth in the era even more distant than geologically recorded time, he 'should expect to be a witness of the evolution of living protoplasm from not living matter'.[56] In any case, Stokes obviously interpreted Thomson as agreeing with him that God created life on the earth. And, in fact, when Kelvin broached the subject in 1897, he mentioned nothing of meteors:

> Mathematics and dynamics fail us when we contemplate the earth, fitted for life but lifeless, and try to imagine the commencement of life upon it. This certainly did not take place by any action of chemistry, or electricity, or crystalline grouping of molecules under the influence of force, or by any possible kind of fortuitous concourse of atoms. We must pause, face to face with the mystery and miracle of the creation of living creatures.[57]

More consistent were his pronouncements on evolution and natural selection. As in the 1871 address, again in 1898 he wrote that although 'the limitation of geological time' undermined the 'potency' of natural selection, evolution was a 'larger question', because 'we cannot put any limit to the rapidity with which evolution may have taken place, so far as there has been evolution at all'.[58]

Stokes argued that 'the progress of science . . . leaves barriers which it gives no indication that science will ever be able to get over; nay, sometimes it makes the existence of such barriers more apparent'.[59] Such barriers were the origin of life on the cooling earth and the gaps between widely different forms of animals whose fossils were preserved in geological strata. Stokes saw 'no prospect' of explaining these by the operation of natural causes.[60] Indeed, respecting the origin of life, 'several persons who, in other respects, go in completely for evolution, allow that, in this case, something more is required'.[61] Here, science provided evidence for God's intervention in nature, an intervention occurring 'not merely at some indefinitely remote time which we please to contemplate as that of the origin of things',[62] but at a time recent enough to be accessible to our understanding. And if in this instance 'we must have recourse to some ultra-scientific cause, there is nothing unphilosophical in the supposition that this ultra-scientific cause may have acted subsequently also'.[63] Subsequent acts included the origin of new sorts of animals, for detectable evolutionary processes were quite limited. Darwin's pigeons were still pigeons, his finches still finches. Clearly, such restricted transmutations as these were inadequate to 'bridge over the enormous interval which separates an oyster from a man', and claims to the contrary were merely 'utterly rampant' speculation.[64] Once again, there was 'general rough accordance' between science and Scripture, because, as Stokes stated in 1864, Genesis's 'grand picture' portrayed God's creation as a succession of creative acts.

The most important of these acts, of course, was the creation of man. Stokes, though still avoiding a *slavish* literalism, insisted on a Genesis interpreted more literally than in other cases:

> In the account of the creation it is distinctly stated that man was separately created, "in the image of God," whatever that may imply. Nor is this a point in which by a wide licence of interpretation we might say the language was merely figurative; that we can afford to understand it so, for that Scripture was not given to teach us Science.[65]

In addition, Biblical discussion of man's origin and initial condition was not confined to Genesis. 'They are dwelt on at length, in connexion with the scheme of redemption, by St. Paul, and, are more briefly referred to by our Lord himself, in connexion with the institution of marriage.'[66] Consideration of man's moral faculties reinforced the Biblical narrative. Both internal inspection of one's own moral convictions plus examination of moral convictions in other cultures demonstrated the persistence of an innate sense of right and wrong even though men often behaved immorally. Men had been created innocent, but had fallen. Against such impressive Biblical and moral evidence, evolution could offer 'nothing more than a hypothesis of continuous transmutation, incapable of experimental investigation, and making such demands on our imagination as to stagger at least the uninitiated'.[67] In fact, 'some even strongly pronounced evolutionists would seek something beyond evolution for the origin of man upon earth'.[68] If man were formed from animals by natural evolutionary processes, then it would be the Author of those processes, not man, who was responsible for man's sinful state. For such reasons, even *stronger* evidence should be required for a theory of human evolution than for other scientific theories:

> If some conclusion to which science seems to point throws a serious difficulty in the way of what we have been in the habit of considering was revealed to us, specially if it be a difficulty of a moral nature, we have a perfect right to demand severer evidence before we can accept it than what might have sufficed to lead us to regard it as in all probability true had there been no such appearance of opposition.[69]

Unsurprisingly, it was Scripture, not science, which gave Stokes a way to allow for the possible verification of human evolution. In 1889, by which time evolution was becoming widely accepted, a Reverend J H Lamb sent Stokes a copy of his paper illustrating 'evolution with Divine intervention by the Scriptural account of the Incarnation', which, Stokes said, 'struck me very much, and is I think calculated to do much good to some whose faith may be tottering in consequence of the supposed demands which science makes for the acceptance of evolution'.[70] In the Incarnation, 'the first inception of the human life of the Incarnate Word was by a supernatural act of Divine power', but thereafter that life developed in an ordinary human way in the womb through childhood into adulthood. 'Here then is an instance of Divine Interposition, gigantic in its result, and yet of such a nature as not to offer any visible exception to the doctrine of evolution: of something superadded to, rather than running counter to, evolution.'[71] Citing Lamb's argument in 1892, Stokes declared that the assumption 'that man took his origin by a supernatural modification of some previously existing animal organism' was 'a position which no study of evolution can overturn, or could overturn, unless we were prepared to account for the whole of the cosmos by mere natural causes, without the intervention of a Supreme Being'.[72] This supposition as much required 'a creative power as if man had been formed directly from materials not endowed with life'.[73]

According to Stokes, therefore, Scripture, science, and morality combined to disclose a God who intervened in the course of nature in various ways. He could act by fiat, leaving empirically detectable traces in the form of discontinuities in the universe's otherwise continuous development. The origin of life was the most obvious such discontinuity, but the origin of species and the origin of man were others. God could also act by fiat, leaving no empirically detectable traces. The Incarnation was the most important example, but if biologists could establish a convincing theory of human evolution, then the origin of man would become another. Third, God could act through established laws. He could, for example, prevent the heat death of the universe suggested by the second law of thermodynamics, merely by utilising natural laws to effect the statistically improbable, but possible, reversal of the dissipation of energy.

Opposed to Stokes's position was not the idea of accepting evolution as a working hypothesis to see where it would lead, but the militant contention that evolution was firmly established, and, more important, that it had already led to materialism. 'Of course', Stokes explained, 'we may assume evolution *for trial*, . . . but to adhere to it when there appears not the slightest prospect of its competence to account for the phenomenon presented does, I confess, seem to me to indicate an *animus* in the direction of endeavouring to dispense with a Creator.'[74] It was one thing to trace evolutionary processes as far as possible, quite another to assume that they 'must' suffice to account totally for the development of the earth and its life. Such a continuous connection between matter, animals and man would render man not just another animal, which was bad enough, but a machine, thus doing away with morality and immortality.

Physics versus biology

Astronomy was the model science for early Victorians. Newton had explained astronomical phenomena with a simple, dynamical theory cast in exact mathematical form. Herschel referred to astronomy as the 'most perfect of sciences'[75] and to the law of gravity as 'the most universal truth at which human reason has yet arrived'.[76] Whewell concluded that 'astronomy . . . is not only the queen of the sciences, but, in a stricter sense of the term, the only perfect science;—the only branch of human knowledge in which particulars are completely subjugated to generals, effects to causes . . .'.[77] Analysed mathematically in terms of Newton's laws of motion, astronomy lay at one end of that spectrum of sciences discussed in Chapter 2. The other areas of science were more or less removed from astronomy depending on how closely theories in each branch resembled those in astronomy. Reflecting this same view and the great progress gained in the physical sciences by 1870, James Clerk Maxwell wrote of chemistry as residing at the far end of the spectrum of the physical sciences, adjacent to the biological sciences. Though dynamics was, according to Maxwell, 'reclaiming large tracts of good ground from one side of chemistry, chemistry is extending with still greater rapidity on the other side into regions where the dynamics of the present day must put her hand upon her mouth'.[78] Or, as Thomson graphically put it in proposing to Stokes that *naturalist* be defined to encompass all those who study nature:

> I hope . . . *naturalist*, will not be objected to, fatally. I know that pigeon fanciers & butterfly & beetle collectors will be desperately offended at being classed with Newton & Faraday, but still I think propriety & convenience of language renders it necessary to disregard their feelings.[79]

The superiority of the exact physical sciences was assumed by Thomson in arguing against uniformitarian geology. In his debate with Huxley in the late 1860s he praised earlier geologists for advancing their science by trying 'to bring geology within the region of physical science, to emancipate it from the dictation of authority and from dogmatic hypotheses'.[80] Thomson thought that geology ought to be regarded as 'a branch of physical science' which sought to accommodate experimental and mathematical investigations and the findings of astronomy and thermodynamics.[81] The trouble with geology in the 1860s was that the prevailing uniformitarianism 'is in direct opposition to the principles of natural philosophy', conflicting with both thermodynamics and modern physical astronomy.[82] The 'very root of the evil' was geologists' disdain for physics as 'something quite foreign to their ordinary pursuits'.[83] Quite naturally Thomson gave little attention or credence to the opposite possibility—that the *physical* arguments must somehow be wrong because they disagreed with detailed geological research, about which Thomson was not all that knowledgeable.[84]

Even more explicit than Thomson, Stokes also argued the point at a deeper level than Thomson had. Contrasting current knowledge of biological and physical laws, for example, Stokes supposed 'that biologists, as well as physicists, would allow that we know more about physics than we do about biology . . .'.[85] Not only was physics more advanced than biology, but it had higher standards governing inferences made from available evidence: '[Darwin's] theory has been accepted by many eminent biologists with a readiness that is puzzling to an outsider, especially one accustomed to the severe demands for evidence that are required in the physical sciences'.[86] When scrutinised by the standards of advanced modern physics, biological evidence was revealed as only adequate to support limited evolution, far short of the extreme, materialistic version being claimed. Such an extreme view could be accepted only in spite of, not because of, biological considerations and only in the absence of legitimate religious considerations. However, as we saw in Stokes's discussion of conditional immortality, advanced modern physics disclosed a world in harmony with the non-materialism of the Bible. While accepting geology's claims against the Bible and rejecting slavish literalism, Stokes fixed his moral and intellectual faculties upon the honest narrative of Jesus's character and teaching. Thus approached, Scripture joined with the methods and content of physics to define biology's present, and future, limitations.

Thus, Thomson's and Stokes's views especially conflicted with John Tyndall's use of physics in conjunction with evolution as part of his 'scientific materialism'. As indicated in Chapter 1, Tyndall's Belfast Address of 1874 buttressed evolutionary theory with the law of the conservation of energy. Because 'living' things were subject to energy conservation the same as non-living, the law helped Tyndall eradicate the distinction between life and non-life. 'Of still wider grasp and more radical significance' even than Darwin's theory, the conservation of energy, according to Tyndall in his Belfast Address,

"binds nature fast in fate" to an extent not hitherto recognized, exacting from every antecedent its equivalent consequent, from every consequent its equivalent antecedent, and bringing vital as well as physical phenomena under the dominion of that law of causal connexion which, so far as the human understanding has yet pierced, asserts itself everywhere in nature.[87]

Thomson had expressed a different view in the early 1850s and continued to do so until the end of his life. In an 1852 paper, he discussed the living organism's transformation of chemical energy into heat and mechanical effect, probably through the intermediary of electrical energy. Though the overall process was governed by the principle of energy conservation, the details of the means of transformation were unknown.

> Whatever be the nature of these means, consciousness teaches every individual that they are, to some extent, subject to the direction of his will. It appears, therefore, that animated creatures have the power of immediately applying, to certain moving particles of matter within their bodies, forces by which the motions of these particles are directed to produce desired mechanical effects.[88]

Or, as he told his natural philosophy class in the spring of 1853, since life did not add energy, 'the [vital] power is a mere power of direction'.[89] In 1862 Thomson and P G Tait published a popular article in *Good Words* on 'Energy' in which they stated: 'It seems even probable that it is actually through electric force that the energy of the food is placed at the disposal of that most inscrutable of finite, created, and subject agencies, a free will directing the motions of matter in a living animal.'[90] Late in life, Kelvin thought that biologists were finally returning to his point of view: 'Modern biologists are coming, I believe, once more to a firm acceptance of something beyond mere gravitational, chemical, and physical forces; and that unknown thing is a vital principle.'[91] In a letter to a correspondent in the year before his death, Kelvin, though without referring to energy considerations, drew the contrast between materialism and a religious notion of free will: 'The perception of every one of the human race of his own individuality and free will seems to me to absolutely disprove all materialistic doctrines and to give us scientific ground for believing in the Creator of the Univ[erse] in whom we live and move and have our being.'[92]

While Thomson made occasional statements in private and in public, Stokes responded more directly and fully to Tyndall than Thomson had, though along the same line as Thomson. As seen in Chapter 1, Stokes used his presidential address to the British Association in 1869 to answer the materialism of Tyndall's address to section A the year before. Writing in 1879, Stokes thought that

Tyndall's 'bold' and 'unflinching' Belfast Address might turn out actually to aid the cause of religion.

> In the attempt to deduce ourselves and our surroundings from that primeval condition of matter by mere evolution, by which I mean the blind operation of natural laws, he is obliged to endow with emotion the ultimate molecules of matter in a fiery nebula, and to adopt a series of conjectures against which common sense rebels. The glove is boldly taken up, and the result is a *reductio ad absurdum*.[93]

Tyndall's materialism may have led to a *reductio ad absurdum*, but it still warranted a specific alternative. Stokes provided one with 'directionism'. It was an alternative Stokes outlined as early as 1879 and finally labelled in 1893. 'To save circumlocution, I will coin a word, and call the view which I have been endeavouring to put before you *directionism*. The alternative views would then be named materialism and directionism.'[94]

Disagreeing with Tyndall, Stokes explained that religious men should not worry that the law of the conservation of energy appeared to hold for living things as well as non-living. The essence of directionism lay in that non-material aspect of man which directed physical activity in the body without opposing it or holding it in check. In contrast to that earlier kind of vitalism, Stokes envisioned 'a directing power, not counteracting the action of the physical forces, but guiding them into a determined channel'.[95] He used the analogy of a moving train, declaring that the human will was like 'the intelligence of the engine-driver' not 'the coals under the boiler'.[96] Indeed, the fact that will appeared to lie outside the arena of energy conversions and transfers caused Stokes also to reject the conservation of energy as providing a relevant argument for immortality.[97]

Directionism obviously dovetailed with Stokes's Biblically derived theories of conditional immortality and the tripartite nature of man, though the latter's notion of potentially immortal ego-spirit was more refined and definite than directionism's notion of guiding will. Moreover, by bringing before us the concept of one entity acting on another *entirely different*, directionism 'led on to the contemplation of that mystery of mysteries, Will, . . . to the contemplation of Will, and of the effects of its exercise'.[98] Just as will influenced body, so also God did nature, both by independent fiat and through established law. Furthermore, directionism resembled Stokes's scientific view, discussed in Chapter 7, of how non-mechanical gravity and ether operated on ponderable matter. In fact, Stokes preferred directionism to other forms of non-materialism because 'that form of the alternative to materialism strikes me as being more nearly analogous to what we know in science than are certain other forms'.[99] Stokes did not think

that natural science could demonstrate the reality of human immortality, but 'I do think that natural science can, by pointing out the
insufficiency of the materialistic hypothesis, remove the apparent
incredibility of any such revival [*sic*, probably should be survival], so
as to leave the mind open to weigh any evidence in favour of survival
that may come from a totally different quarter.'[100] Consequently,
directionism, with its analogical alliance with natural science, was
strategically placed to counter materialism and its attendants, extreme
evolution and scientific naturalism.

Whereas Tyndall joined physics and biology against theism, Stokes
and Thomson fused physics and theism against biology. In effect,
Thomson and Stokes used physics to help adjudicate disputes between
theology, on the one hand, and biology and geology on the other. They
did so, however, in different ways. Thomson did not marshal theological or Biblical arguments directly against geological uniformitarianism, but his physical arguments against uniformitarianism did
have the important theological consequence of defending the design
argument against Darwinian natural selection. Stokes did directly
employ both the design argument and Biblical evidence—as reinforced
by considerations of physics—to help deny the occurrence of evolution
itself. Part of the conflict involved in the whole Victorian discussion of
science and religion, therefore, was conflict between areas of science.
Thomson and Stokes, giving special place to physics, were finding
their ways through an age of controversy by combining religion and
science as well as possible from their respective 'centrist' and
'conservative' perspectives.

1   Michael Ruse 1975 'The Relationship between Science and Religion in
    Britain, 1830–1870' *Church History* **44** 505–22. For an extended
    discussion of the inadequacies of the conflict view of science and
    religion, see J R Moore 1979 *The Post-Darwinian Controversies: A Study
    of the Protestant Struggle to Come to Terms with Darwin in Great Britain
    and America, 1870–1900* (Cambridge: Cambridge University Press)
    pp. 19–122.
2   H P Stokes 1903 'Reminiscences of Sir George Stokes' *The Cambridge
    Chronicle* (13 February). I have used the clipping in the Stokes Papers at
    Pembroke College, Cambridge. For brief discussions of Anglican
    Evangelicalism see B M G Reardon 1971 *From Coleridge to Gore:
    A Century of Religious Thought in Britain* (London: Longman) pp. 23–31
    and the 'Introductory Essay' in E Jay (ed) 1983 *The Evangelical and
    Oxford Movements* (Cambridge: Cambridge University Press).
3   For example, he wrote that polygamists converted to Christianity
    should be allowed to remain polygamists. (Stokes, 'On Polygamy in

Connection with Christian Missions' reprint from *The Churchman* contained in *CUL*.)

4   William Paley 1834 *A View of the Evidences of Christianity* in *The Works of William Paley* (London: Jennings and Chaplin) p. 6.

5   *Ibid*. p. 6.

6   Charles Darwin 1969 *The Autobiography of Charles Darwin, 1809–1882* ed Nora Barlow (New York: W W Norton & Co.) p. 59.

7   William Whewell *On the Foundations of Morals* four sermons preached before the University of Cambridge in 1837 (Cambridge and London, no date). Stokes's notes on the sermons are in *CUL Stokes Collection* PA36.

8   Stokes *Natural Theology* I 227–8.

9   Stokes's remarks made in the discussion following his address 'On the Bearings of the Study of Natural Science, and of the Contemplation of the Discoveries to which that Study Leads, on Our Religious Ideas' *Journal of the Transactions of the Victoria Institute* **14** (1880) 227–38. The quotation is on p. 247.

10 Stokes *Natural Theology* I 24.

11 Stokes 1897 *Conditional Immortality: A Help to Sceptics* (London: James Nisbet) p. 81.

12 Stokes to A H Tabrum, 5 October 1899 *Memoir* I 80. See Stokes *Natural Theology* II 215–22, where he discussed Christ's character based on the gospels 'regarded as simple history'.

13 E King 1909 *Lord Kelvin's Early Home* ed E T King (London: Macmillan) p. 49.

14 J Thomson to R Douglas, 3 March 1849, in J Thomson 1912 *Collected Papers in Physics and Engineering* ed J Larmor and J Thomson (Cambridge: Cambridge University Press) p. xl.

15 J Thomson to R Douglas, 15 August 1846, in Thomson *Collected Papers* (note 14) p. xxxvii.

16 Notes on Fleming's lectures *CUL Kelvin Collection* NB21.

17 Notes on Meikleham's lectures, 9 November 1839 *CUL Kelvin Collection* NB9.

18 J P Nichol 1838 *The Phenomena and Order of the Solar System* (Edinburgh and London) p. 120.

19 *Ibid*. pp. 64–5.

20 J Thomson to R Douglas, 3 March 1849, in Thomson *Collected Papers* p. xl; *Thompson* II 1087–8.

21 Note dated 1 March 1851 on Thomson's draft of 'On the Dynamical Theory of Heat, with Numerical Results Deduced from Mr Joule's Equivalent of a Thermal Unit, and M. Regnault's Observations on Steam' *CUL Kelvin Collection* PA128. The note has been published in D B Wilson 1974 'Kelvin's Scientific Realism: The Theological Context' *The Philosophical Journal* **11** 58–9. Thomson's paper (without the note) is in *MPP* I 174–210.

22 Thomson 'Presidential Address to the British Association, Edinburgh, 1871' *PLA* II 205.

23 A G King 1925 *Kelvin the Man: A Biographical Sketch by His Niece* (London: Hodder and Stoughton) pp. 28–9.

24   Thomson to P G Tait, 6 April 1890, Edinburgh University Library.
25   Stokes *Conditional Immortality* pp. 28–9.
26   Darwin *Autobiography* p. 87.
27   S Budd 1967 'The Loss of Faith: Reasons for Unbelief among Members of the Secular Movement in England, 1850–1950' *Past and Present* no 36 118.
28   G Rowell 1974 *Hell and the Victorians: A Study of Nineteenth-Century Theological Controversies Concerning Eternal Punishment and the Future Life* (Oxford: Clarendon) p. 212.
29   Stokes 1882 *Evidence of Missionaries as to the Practical Effect of Presenting Christianity to the Heathen in the Form Associated with the Doctrine of "Life in Christ"* (Cambridge: privately printed).
30   Stokes's copy of his letter to Mr Barton, 12 December 1881 *CUL Stokes Collection* B163. Relevant letters on the controversy are scattered through many parts of the collection.
31   E White 1878 *Life in Christ: A Study of the Scripture Doctrine on the Nature of Man, the Object of the Divine Incarnation, and the Conditions of Human Immortality* 3rd edn (London: Elliot Stock) p. iii. This is the third edition of White's book, published first in 1875, which White regarded as a different book, though with the same basic title, from the one published in 1846. Some 265 letters from White to Stokes are in *CUL Stokes Collection*.
32   H Constable 1873 *Hades; or the Intermediate State of Man* (London: Elliot Stock) p. 8.
33   Stokes *Natural Theology* II 34.
34   See Stokes 1890 '"I": A Lecture on the Immortality of the Soul', *The Family Churchman* (9 April) 1–23, and 'Is the Soul of Man by Its Nature Immortal?', a paper read at Sion College on 20 April 1893, a reprint of which is in *CUL*.
35   Stokes *Conditional Immortality* p. 19.
36   Stokes *Natural Theology* I 95.
37   *Ibid.* I 97.
38   *Ibid.* I 102.
39   E J Routh to Stokes, 2 November 1902 *CUL Stokes Collection* R1118. The notes by Stokes are undated but obviously date from the last few months of his life, as he died in February 1903.
40   Stokes '"I"' p. 14. He also discussed there Romans 8:10, one of the versus listed on the back of Routh's letter.
41   Stokes '"I"' p. 15. Stokes gave as examples Acts 7:59 and Hebrews 12:23.
42   This verse is listed on the back of Routh's letter.
43   Stokes '"I"' p. 17. Stokes quoted from 1st Corinthians 15:35–37.
44   *Ibid.* p. 21.
45   *Ibid.* p. 21.
46   Stokes to C E Grove, 28 July and 8 September 1864, *CUL* Add. MS 5989, sheets 45, 46, 55, and 56.
47   Stokes 1891 'Genesis and Science' *The Expositor* 4th series **3** 46.
48   Stokes to A H Tabrum, 17 August 1900 *Memoir* I 86–7.

49 Stokes 'Genesis and Science' p. 51.
50 Thomson 'On Mechanical Antecedents of Motion, Heat, and Light' *MPP* II 38. For Whewell's justification of the view that God must have intervened frequently in the history of the earth's life, see W F Cannon 1960 'The Problem of Miracles in the 1830's' *Victorian Studies* **4** 5–32.
51 Thomson 'Presidential Address' (note 22) p. 203. See also Thomson 'On Geological Time' *PLA* II 64 [1868].
52 Thomson 'Presidential Address' p. 204.
53 *Ibid.* p. 204.
54 *Ibid.* p. 204. My italics.
55 Stokes 1893 'Science and Faith' *Official Report of the Church Congress* **33** 343. My italics.
56 T H Huxley 1870 Presidential address *Br. Assoc. Rep.* p. lxxxiv. Emphasising throughout this part of the lecture the absence of evidence supporting his conclusion, Huxley declared that 'I beg you once more to recollect that I have no right to call my opinion any thing but an act of philosophical faith.' (*Ibid.* p. lxxxiv.)
57 Kelvin 'The Age of the Earth as an Abode Fitted for Life' *MPP* V 230. Kelvin gave this address at Stokes's request to the Victoria Institute in 1897.
58 Kelvin to E Davys, 5 October 1898 *GUL Kelvin Papers* LB5/159.
59 Stokes 'On the Bearings of the Study of Natural Science' (note 9) p. 232. See Stokes 1889 'Literature of the Day, and Its Attitude towards Christianity' *Official Report of the Church Congress* **29** 212.
60 Stokes 'On the Bearings of the Study of Natural Science' pp. 235–6.
61 Stokes *Natural Theology* II 181. Here, as in the case of the origin of man below, Stokes may well have been referring to A R Wallace, who independently of Darwin formulated the theory of evolution by natural selection but who, in certain respects, broke with Darwin in the 1860s. Through a combination of psychical research and the study of primitive man, Wallace decided that an 'overruling intelligence' (not the Christian God) existed and that it had intervened at specific points in the world's otherwise continuous development. Two such points were the origin of life and the origin of mankind. (See, for example, F M Turner 1974 *Between Science and Religion: The Reaction to Scientific Naturalism in Late Victorian England* (New Haven: Yale University Press) pp. 68–103.)
62 Stokes *Burnett Lectures* p. 331.
63 *Ibid.* p. 335.
64 Stokes 1879 'Religious Benefits from Recent Science and Research' *Official Report of the Church Congress* **19** 424–5.
65 Stokes 1883 'On the Absence of Real Opposition between Science and Revelation' *Journal of the Transactions of the Victoria Institute* **17** 200.
66 *Ibid.* p. 201. See Stokes 'Science and Faith' p. 344.
67 Stokes 'On the Absence of Real Opposition between Science and Revelation' p. 201. See also, Stokes *Natural Theology* I 224–33; II 181–2, 197–8.
68 Stokes *Natural Theology* II 166.

69   Stokes 'On the Absence of Real Opposition between Science and Revelation' p. 204.

70   Stokes's copy of his letter to Rev. J S Exell editor of *The Homiletic Magazine*, 17 April 1889 *CUL Stokes Collection* E241.

71   *Ibid.*

72   Stokes *Conditional Immortality* pp. 43–4.

73   Stokes *Natural Theology* II 167. See Stokes to A H Tabrum, 7 August 1900 *Memoir* I 85.

74   Stokes *Natural Theology* I 170–1. See *ibid.* II 240.

75   J F W Herschel 1966 *A Preliminary Discourse on the Study of Natural Philosophy* a facsimile of the 1830 edition (New York: Johnson Reprint Corp.) §67.

76   *Ibid.* §116.

77   W Whewell, Address *Br. Assoc. Rep.* (1833) p. xiii.

78   J C Maxwell 1885 'Physical Sciences' *Encyclopaedia Britannica* 9th edn (Edinburgh: Adam and Charles Black) XIX 3.

79   Thomson to Stokes, 14 April 1862 *CUL Stokes Collection* K135.

80   Thomson 'On Geological Time' (note 51) p. 10.

81   Thomson 'Of Geological Dynamics' *PLA* II 113 [1869].

82   Thomson 'On Geological Time' p. 44; 'Of Geological Dynamics' pp. 108–9.

83   Thomson 'Of Geological Dynamics' pp. 74–5.

84   In supporting Thomson's position, his friend and collaborator P G Tait unequivocally declared the superiority of the exact sciences: 'Let us then hear no more nonsense about the interference of mathematicians in matters with which they have no concern; rather let them be lauded for condescending from their proud preeminence to help out of a rut the too ponderous waggon of some scientific brother.' (Quoted in J D Burchfield 1975 *Lord Kelvin and the Age of the Earth* (New York: Science History Publications) p. 93.)

85   Stokes *Natural Theology* II 236–7.

86   Stokes 'On the Absence of Real Opposition between Science and Revelation' (note 65) p. 200. He made the same point in 'On the Bearings of the Study of Natural Science' (note 9) p. 235.

87   J Tyndall, Presidential address *Br. Assoc. Rep.* (1874) p. lxxxviii. 'Scientific Materialism' is the title Tyndall gave to the reprint of his 1868 presidential address to section A in his *Fragments of Science* 2 vols (New York: P F Collier & Son, 1902) II 82–98. His more famous address of 1874 was reprinted simply as 'The Belfast Address' in *Fragments of Science* II 145–214. Tyndall seems not actually to have been, strictly speaking, a materialist, though his language and manner invited that interpretation of his views. (See D B Wilson 1984 'A Physicist's Alternative to Materialism: The Religious Thought of George Gabriel Stokes' *Victorian Studies* **28** 69 and J D Burchfield 'Empiricism and Idealism in the Science of John Tyndall' unpublished. I am grateful to Professor Burchfield for allowing me to read his paper.)

88   Thomson 'On the Mechanical Action of Radiant Heat or Light: On the Power of Animated Creatures over Matter: On the Sources Available

to Man for the Production of Mechanical Effect' *MPP* I 509. Thomson wrote to his brother in 1862 that he did not think that vital processes violated either the first or second law of thermodynamics. (Thomson to J Thomson, 21 April 1862, in J Thomson *Collected Papers* (note 14) pp. lvii–lviii.)

89   W Jack, Notes on Thomson's lectures for 1852–53 *GUL* MS Gen 130. In 1862 he told the class: 'The difference between a living agent and a piece of dead matter is in that wonderful fact of the living being directing the force.' (D Murray, Notes on Thomson's lectures for 1862–63, 19 November 1862 *GUL MS Murray* 325.)

90   Thomson and P G Tait 1862 'Energy' *Good Words* 605.

91   'Lord Kelvin on Science and Theism' *The Nineteenth Century* **53** (1903) 1069.

92   Kelvin to Professor J Helder, 12 May 1906 *CUL Kelvin Collection* LB31.2, printed in Wilson 'Kelvin's Scientific Realism' (note 21) p. 60.

93   Stokes 'Religious Benefits from Recent Science and Research' p. 424. He felt the same two decades later: 'Tyndall . . . was led to attribute emotion to the ultimate molecules of matter in a fiery mass of gas!' Stokes to A H Tabrum, 3 August 1900 *Memoir* I 83.

94   Stokes *Natural Theology* II 47. The 1879 reference is Stokes 'Religious Benefits from Recent Science and Research' pp. 421–2. He repeated the idea in 1880 in 'On the Bearings of the Study of Natural Science' p. 230.

95   Stokes *Natural Theology* II 46.

96   Stokes 'On the Bearings of the Study of Natural Science' p. 230.

97   Stokes *Conditional Immortality* pp. 32–4. See Rowell *Hell and the Victorians* (note 28) p. 205.

98   Stokes *Natural Theology* II 47, 54.

99   *Ibid.* II 55.

100 *Ibid.* II 57. Though Stokes and Thomson may have been the best-known physicists opposing Tyndall, they were not the only ones. The others included Balfour Stewart, P G Tait and Oliver Lodge. (D B Wilson 'A Physicist's Alternative to Materialism' pp. 94–6.)

5

Cautious Realism

'Cautious realism' is not an expression used by Stokes and Thomson. Nevertheless, I think it provides an accurate description of their methodological position in regard to man's knowledge of unobservable physical entities. *Realism* means that they thought actual knowledge of actually existing unobservables was possible. *Cautious* refers to their conviction that such knowledge was difficult to attain and, therefore, that one should be exceedingly careful not to claim to know too much. In cases where sufficient evidence or insight was absent, one proceeded by avoiding specific assumptions about unobservables or by employing analogical arguments which depended on an essential similarity between the observable realm and the unobservable. The hope, however, was eventually to go beyond these approaches to an account of the hidden realm itself.

The *possibility*, in principle, of knowledge of unobservables and the *proper place* of such entities in any particular theory are related but not identical issues. Accordingly, there are those who have written a great deal about the first issue, but who have not themselves constructed scientific theories. Conversely, there are those, like Stokes and Thomson, deeply engaged in formulating physical theory—and of necessity employing a methodology in the process—but who have written very little about the methodology itself. Stokes and Thomson lacked some of the philosophical subtlety of their younger colleague, Maxwell, who thought at great depth on the connections between physics and philosophy.[1] We must not mistake Stokes's and Thomson's use of unobservables in any particular instance as necessarily representing their general methodological position. Rather, one looks for methodological statements, however brief, and tries to correlate them with evidence from a range of physical theories formulated by the two. In addition, one examines the historical context of their formative years for methodological viewpoints they may have been following. Stokes and Thomson's similar methodology of cautious realism was a flexible position allowing diverse approaches in specific instances, and, accordingly, they could certainly differ from one another in emphasis from time to time.

This chapter discusses their early careers, in the conviction that their early views persisted, providing the framework for their later thoughts about such things as ether, cathode rays and radioactivity. The first section considers the early-Victorian background to their views, while the next two discuss Stokes and Thomson themselves. The chapter concludes by analysing the theological context of their cautious realism.

Early-Victorian background

This section focuses on the methodological views of principal figures associated with early-Victorian Glasgow and Cambridge, men with whom the young Thomson and Stokes were in fairly direct contact. It first surveys the earlier views of Newton, the Scottish 'commonsense' philosophers and early-nineteenth-century French physicists.

Newton's writings provided a powerful legacy, which could be construed in various ways. In his famous statement, 'hypotheses non fingo', he declared that he would not feign hypotheses about the cause of gravity. This could be taken to mean that one should not speculate on underlying causes at all, or that one should only do so with good evidence. The first of his four 'rules of reasoning' stated that one should only employ 'true causes', and only as many as required to explain the phenomena. The third rule concluded that the *unchanging* qualities of observable bodies 'are to be esteemed the universal qualities of all bodies whatsoever'. For Newton the third rule affirmed, for example, the fundamental sameness of the macroscopic and microscopic realms, thereby guaranteeing that, by 'the analogy of nature', the smallest particles of matter were extended and impenetrable.[2] The third rule, or at least the viewpoint it stood for, informed the views of those early-Victorian natural philosophers who thought that the microscopic world, like the macroscopic, was mechanical and that, therefore, mechanical models could, by analogy, provide insights into the structure and processes of nature's hidden realm.

Beginning with Thomas Reid (1710–96), the Scottish commonsense school emphasised the Newtonian dictum against hypotheses. Though he may not have absolutely dismissed the possibility of knowing unobservables,[3] Reid's main thrust was to limit 'causes' in natural philosophy to descriptive laws. Underlying 'efficient' causes did exist—laws described the results of their activity—but they were impossible, or at least enormously difficult, to know. The view marked Reid's acceptance of much of David Hume's argument that one can only know the 'constant conjunction' of events, but without what Reid saw as Hume's unrelieved scepticism. Events which for Hume were only observed to succeed one another—and nothing more—were for Reid the embodiment of regular laws created by God which one could discover because of an inherent drive to do so. For Reid, Newton's first-rule requirement for *true* causes weighed against use of hypothetical entities and analogies between different fields. The first could not be known actually to exist because they were unobservable, and the second could mislead. Later commonsense philosophers, Dugald Stewart (1753–1828) and Thomas Brown (1778–1820), loosened Reid's strictures against hypotheses and analogies. They could be *useful* in helping one to remember a great deal of factual information or *fruitful* in suggesting further laws. But they did not portray a real, hidden realm. Stewart even accepted Boscovich's theory of matter in this vein.[4]

French physicists also pursued Newtonian themes. Emphasising Newton's success in astronomy, the Laplacian tradition perceived the physical world in terms of particles of matter and action-at-a-distance forces, with gravitational theory as the best example. Light consisted of corpuscles of matter which interacted through forces with ordinary matter. Heat, electricity and magnetism each involved an imponderable fluid, which itself was composed of particles acting on one another and on particles of ordinary matter through forces. This highly realist approach to the unobservable was challenged by Fourier in heat and Ampère in electricity and magnetism, who, like commonsense philosophers, thought it important not to feign hypotheses. Assimilating this latter tradition into his philosophy of science, which excluded unobservables from physical theory, the positivist philosopher August Comte established an important French point of view. It has been argued that the influence of Comtean positivism contributed to the decline of French physics in the middle of the century.[5] Although Fresnel's wave theory of light challenged aspects of the Laplacian tradition (by rejecting light corpuscles and by trying to unify phenomena into one ethereal medium), it was still Laplacian in its conception of an ether constituted of particles acting on one another through action-at-a-distance forces. Developed further by Cauchy, this conception of the ether influenced British theorists, especially at Cambridge.[6]

Such is the background to British thought of the 1830s and 1840s. Let us now turn to three Glasgow professors—Thomas Thomson, Meikleham, and Nichol—and several men at Cambridge.

Thomas Thomson presented a cautious view of man's current knowledge of electricity and heat. He asked at one point about the cause of electrical properties induced in bodies by friction, answering: 'We are altogether ignorant of it; though a variety of hypothetical explanations have been advanced.' These hypotheses proposed the existence of zero, one or two electrical fluids. 'Whatever the cause of these new properties may be', Thomson wrote, 'we may distinguish it, although unknown, by applying to it the term *electricity*.'[7] In a chapter 'On the nature of heat', Thomson said that 'philosophers are at present divided in their sentiments on this subject'. Some regarded heat as a 'peculiar substance', others as 'a property of matter, a motion of a particular kind'.[8] One theory afforded a good explanation of some phenomena of heat, the second of others. Thomson did not think, for example, that the substance theory of heat was disproved by Count Rumford's experiments on the amount of heat produced as a cannon was bored. At least part of the heat could have 'proceeded from an augmentation of density, and a consequent diminution of specific heat in the metallic cylinder subjected to friction'.[9] Given such uncertainties: 'I think, therefore, that it will be safest for us, in the present state of our knowledge, to acknowledge our inability to solve this difficult problem, and to confess that we are incompetent to decide whether it be a substance or a quality.'[10] Thomson's was not a methodological proscription of knowledge of unobservables, only a call for extreme care 'in the present state of our knowledge'. Nor did Thomson present Fourier as one who rejected the possibility of such knowledge, but rather as one who investigated the mathematical laws of heat without attempting 'to settle the question respecting the nature of heat'.[11]

Meikleham had presented to classes a critical view of physical knowledge over the years. Terms like *gravitation, attraction* and *inertia*, Meikleham warned, only relate to observable effects and 'convey to us no notion of the real nature of the qualities or agents represented by them'.[12] Likewise, he cautioned William Thomson's class in 1839 that we do not actually know that light, heat, electricity and magnetism are imponderable *substances:* 'We know them, not as substances, but as states of bodies.'[13] Meikleham did not preclude the possibility of knowledge of the unobservable, however. In the 1820s, for example, he explained friction in terms of the microscopic features of surfaces rubbing together and proposed a particular theory to explain the reflection of particles of light.[14]

Nichol could also sound cautious in his writings and lectures. In his popular books on astronomy he seemed to regard laws as the goal of scientific knowledge. In his lectures in 1839–40 to the natural philosophy class, he defined science not as a search for underlying causes but for *relationships*. 'The object of science is the discovery of relationships by means of which the complex may be deduced from the simple.'

Nichol credited Oersted and Ampère with discovering the 'relation' between electricity and magnetism. Thomson's notes on Nichol's extended discussion of the wave theory of light do not mention the ether and its properties. Nichol did think, however, that Ampère's theory that magnetism was produced by electric currents had been rendered 'probable' through its successful predictions, but that it remained 'theoretical'.[15] It may indeed have been Boscovich's theory that, in Nichol's opinion, was science's most penetrating insight into nature's hidden realm. We can supplement Thomson's obviously incomplete record of Nichol's remarks on the subject (see figure 2.2) with Nichol's own text for a popular lecture he had given in Montrose a decade before. There he declared:

> But facts are not sufficient for his [i.e. man's] unwearied Reason, nor does it seem that the mere record of what *is*, wonderful as that may be, is half so enticing to him as the speculation respecting the mystery of the constitution of all things. Accordingly in every age it has been the principal philosophic Inquiry, not so much how portions of nature appear to the eye, or what are the rules which guide them through their cycle of variations, as what causes them so to appear—by what concealed influence they are upheld and what in reality is their mode of Being.

There followed a claim for depriving matter of its 'dead and crude character' and referring it, instead, 'to the constant play of mighty forces'. Nor, Nichol argued, should one confuse Boscovich's 'truly physical' speculation with 'attempts of ingenious men to reduce matter to a mere sensation or high theological idea'. It was 'neither a metaphysical or theological speculation but the result of physical inquiry'.[16]

Given the importance of optics at Cambridge, we should focus on statements there about the luminiferous ether. In his textbook G B Airy distinguished between the 'geometrical' and 'mechanical' parts of the undulatory theory. Though not 'certainly' true as was the geometrical part, the mechanical part—which included suppositions about the 'constitution of the ether'—was 'generally probable'.[17] In the Laplacian manner, Airy treated the ether as composed of particles which acted at a distance on one another. He frequently employed analogies of sound and water waves to explicate the undulatory theory, and when they failed he used the analogy of colliding material bodies.[18] William Hopkins told his students that according to the undulatory theory, 'a perfectly elastic medium pervades all space'. Like Airy, to whose work he frequently referred his students, Hopkins regarded the ether as a system of points.[19] This theory of a 'punctiform' ether guided a good deal of research by Cambridge graduates of the 1830s and 1840s, including Philip Kelland, Samuel Earnshaw and Matthew O'Brien.[20] George Green, on the other hand, preferred not to rely on such detailed accounts of the ether because 'we are so perfectly ignorant of the mode of action of the elements of the luminiferous ether on each other'.[21] James Challis explained in his lectures that 'light is an effect of the action of a very elastic medium, as sound is produced by means of the air'.[22] Also, he thought it might be possible to treat the ether 'en masse', as one did air in acoustics, rather than as a medium constituted of particles.[23] William Whewell's *a priori* philosophy of science declared that man possessed a number of 'fundamental ideas' that constituted a necessary part of his knowledge of nature. In optics the idea of a medium required that there exist some medium to convey light. A 'consilience of inductions' (i.e. the accumulation of evidence from different phenomena) clearly supported the undulatory theory, indicating that the medium in question was the ether, not corpuscles.[24] A strong supporter of the undulatory theory, Whewell speculated that the ether might also be involved in the less well understood phenomena of heat, electricity, and magnetism. He regarded the ether as one of the creations of 'a most wise and good God' which made life possible.[25]

In summary, though the view of the ether in Cambridge was not a straightforward or naive realism, it was a realism not so bound up with the caution displayed by the Glasgow professors. That was probably in part because the undulatory theory was not so well entrenched at Glasgow as at Cambridge. Also, the Glasgow professors were discussing areas of science where theories were less well established than in optics, areas where Cambridge men were more cautious, too. Moreover, the Glasgow professors were undoubtedly more strongly influenced by commonsense philosophy, which had often been directed against ether theories, than were their Cambridge counterparts. Epitomising the difference was the Scot David Brewster's blast at Whewell for 'maintaining that the luminiferous ether really exists, filling universal space, and occupying the pores of all bodies whatsoever'.[26] At the same time, however, the Glasgow professors do appear less severely sceptical than Reid, Stewart and Brown.

Within the variety of early-Victorian thought, probably the most influential writer was the Cambridge graduate John Herschel. To be scientific in this era, Cannon tells us, meant to be like John Herschel.[27] William Thomson, in his introductory lecture at Glasgow, included a long quotation from Herschel's *Preliminary Discourse on the Study of Natural Philosophy* (1830), the only contemporary publication cited in the lecture.[28] Herschel's chapter, 'Of the Higher Degrees of Inductive Generalization, and of the Formation and Verification of Theories', offered guidelines for investigating unsettled branches of natural philosophy. The 'immediate object' was the 'analysis of phenomena, and the knowledge of the hidden processes of nature in their pro- duction', including 'a discovery of the actual structure or mechanism of the universe and its parts, through which, and by which, those processes are executed'. However, since such invisible 'agents' were 'only to be traced by the effects they produce', such knowledge came with exceeding difficulty. Very little was known, for example, about the nature of heat, the nature of an electric current and the cause of gravity. Nevertheless, one must continue 'framing hypotheses and constructing theories' not only because they guided research, but because they could become well enough supported—as with the undulatory theory of light—to yield knowledge about nature's hidden realm.[29] Such successes were so rare in the current state of science, however, that Herschel urged a strategy of *'in the first instance'* de- emphasising discovery of nature's mechanisms in favour of determining

> whether our theory truly represent *all* the facts, and include *all* the laws, to which observation and induction lead. A theory which did this would, no doubt, go a great way to establish any hypothesis of mechanism or structure, which might form an essential part of it: but this is very far from being the case, except in a few limited instances; and, till it is so, to lay any great stress on hypotheses of the kind, except in as much as they serve as a scaffold for the erection of general laws, is to "quite mistake the scaffold for the pile".[30]

In his own writings concerning the unobservable activity of ether and matter, Herschel, like Airy in his *Mathematical Tracts*, had employed macroscopic analogies. He discussed the behaviour of tuning forks in trying to understand the absorption of light, and in considering double refraction he imagined a system of regularly arranged ellipsoidal shells.[31] Herschel emphasised 'the great importance of possessing a stock of analogous instances or phenomena which class themselves with that under consideration, the explanation of one among which may naturally be expected to lead to that of all the rest'.[32]

Clearly, the early-Victorian background offered Stokes and Thomson a variety of viewpoints. Yet, there was a balance of view tending towards their cautious realism. The science professors at Glasgow were less sceptical than the commonsense philosophers, even though under their influence. Green, Airy and Whewell all accepted the ether's reality, but avoided committing themselves to detailed views of its properties. As a close reader of Thomas Brown and a close friend of Airy and Whewell, Herschel in effect presented an amalgam of Scottish caution and Cambridge realism.[33] Herschel's high regard for mathematics and dynamics (see Chapter 2), his cautious approach to nature's hidden processes and structures, his goal of dynamical theories of such mechanisms, and even his analogical use of models combined in accurate definition of Stokes and Thomson's cautious realism. Just as the Glasgow scientific context made it plausible for Thomson to turn to Fourier, for example, the Glasgow–Cambridge methodological context worked against his interpreting Fourier in a strictly positivistic way.

Stokes

This section focuses on three aspects of Stokes's research in the 1840s and 1850s which illustrate his viewpoint. Hydrodynamics, more specifically friction in fluids, his first major area of research, exemplifies the range of his methodology. That work is reinforced by an optical paper of 1849 which contains his longest and most explicit methodological statement. Third, his conjectures on the causes of fluorescence supplement these earlier papers by fully displaying the realist side of his position.

In an unpublished manuscript from the early 1840s, Stokes set his hydrodynamical problem. The generally accepted equations of fluid motion neglected any viscosity that might be present in fluids. After showing that the equations led to specific results at variance with experiment, Stokes concluded: 'But probably fluids are not what they are mathematically supposed to be, and the difference may be the true cause of the discrepancy between theory and observation.'[34] Establishing fluid friction as a true cause thus became the main concern of his hydrodynamical researches. Like Newtonian gravity, fluid friction was to be established as a true underlying cause of fluid phenomena, but in a way that was independent of conjectures concerning even deeper underlying causes—independent, that is, of particular theories of molecular activity. Hence, his first published attack on the problem, in 1843, tried to determine the exact significance of fluid friction by calculating results for frictionless fluids and comparing the results with experimental findings. This seemed 'the only way by which to estimate the extent to which the imperfect fluidity of fluids may modify the laws of their motion, without making any hypothesis as to the molecular constitution of fluids'.[35] In this paper he left even the existence of 'ultimate molecules of fluids' an open question.[36] Not producing a definitive conclusion, the exercise led to a more direct approach to the problem.

In both his unpublished manuscript on fluid friction and his major paper on the subject published in 1845, he did discuss ultimate molecules. In both, Stokes considered two different molecular hypotheses. Both hypotheses relied on the fact that, as he stated in his paper: 'It is an undoubted result of observation that the molecular forces, whether in solids, liquids, or gases, are forces of enormous intensity, but which are sensible at only insensible distances.'[37] He coupled this standard idea with another—that 'there is no doubt that the distance between two adjacent molecules is quite insensible'.[38] It followed that whether the molecules were separated from one another or whether they were smooth particles in actual contact, the implications would be the same.[39]

Specifically, because of the immense number of molecules in a small volume, one was justified in defining 'the velocity of a fluid at any particular point' to mean the velocity common to all the molecules at the point, neglecting the irregular motions they might have with respect to one another. Because of the great intensity of molecular forces, the motion of molecules in an element of fluid with respect to each other would be governed not by weak external forces like gravity, but by molecular forces.[40] If there were no relative motion of the molecules within the element, then as in hydrostatics the pressure would be the same in all directions within it. The equality of pressure in all directions was also the basic assumption underlying the ordinary equations of hydrodynamics. It was equivalent to assuming 'that the mutual action of two adjacent elements of the fluid is normal to the surface which separates them'.[41] Therefore, for a 'tangential action'— that is, friction—to be introduced, there had to be some relative motion of molecules with respect to each other in the element of fluid.

Stokes introduced the required tangential action by discussing molecular activity. In his unpublished manuscript, he formulated his idea of 'starts'. Although one could generally suppose the molecules in positions of equilibrium and, therefore, the pressure equal in all directions,

> in the course of the shifting of the fluid molecules will now and then, here and there start into new positions of equilibrium. Each such start will produce a vibration among the molecules adjacent to the one which started, which will be propagated throughout the mass with a very great velocity, and will be insensible at a small distance from the molecules. While this is going on the pressures about a point where the molecules experience this vibration will not probably be equal in all directions. . . . The effect of the vibration accompanying each start may be considered as the effect of an impulse.[42]

Because of the large number of molecules in a small volume of fluid, starts would occur frequently enough for one to employ a continuous force rather than a 'succession of impulses'. In his published version, Stokes first asked his reader to 'consider what would take place if the fluid consisted of smooth molecules acting on each other by actual contact' and argued that enormous molecular forces would cause molecules to shift suddenly into new positions of equilibrium. And the same thing would happen, he declared, on the more probable view that the molecules were separated from one another.[43]

Despite his discussions of molecules, Stokes pointed out that his conclusions did not depend on molecular theories. He contrasted his work to that of the Frenchman, Navier, who had arrived at similar conclusions. In Navier's treatment of fluid friction, repulsive forces between two molecules depended on the relative motion between them 'so that', as Stokes explained in 1846, 'two molecules repel each other less strongly when they are receding, and more strongly when they are approaching, than they do when they are at rest'.[44] The strength of Stokes's own method, by contrast, was that it 'does not necessarily require the consideration of ultimate molecules'.[45] That is to say, Stokes did not know enough about molecules to support a specific theory about them like Navier's, but he did know enough to know that he could legitimately treat the fluid as a continuum. And because his results did not rest upon a specific molecular hypothesis, they would remain valid whatever the truth of the molecular situation might eventually turn out to be.

Finally, Stokes's proposal of two possible molecular theories— molecules in contact or separated—represents two aspects of his cautious realism. First, the point already made: whichever of these, or other, particular theories might be true, fluid friction was demonstrated to be a true cause. Second, the admittedly unlikely concept of smooth molecules in contact seems to have functioned as a dynamical analogy to the much more probable concept of separated molecules. Years later, Stokes explained his initial thinking to Thomson:

> Suppose you had an Indian rubber bag, in tension, filled with smooth marbles, or at least marbles which were very nearly smooth. Then if you were to give it a motion of distortion, every now and then a marble would start into a new position among its fellows, and the change would be accompanied by a sudden disturbance among the whole set, though strongest in the neighbourhood of the marble that started.[46]

Stokes may not have had this marble-bag model explicitly in mind in the 1840s, but the smooth molecules seem to have played a similar analogical function.

Hence, using a dynamical model and a few fundamental well established conclusions about molecular reality, Stokes formulated the physical basis for demonstrating fluid friction to be a true cause, all the while successfully avoiding unsupportably detailed accounts of molecular activity.

The same blend of caution and realism pervaded Stokes's methodological pronouncement of 1849:

> It may appear to some to be superfluous to deduce particular results from hypotheses of great generality, when these results may be obtained, along with many others which equally agree with observation, from more refined theories which start with more particular hypotheses. And indeed, if the only object of theories were to group together observed facts, or even to allow us to predict the results of observation in cases not very different from those already observed, and grouped together by the theory, such a view might be correct. But theories have a higher aim than this. A well-established theory is not a mere aid to the memory, but it professes to make us acquainted with the real processes of nature in producing observed phenomena. The evidence in favour of a particular theory may become so strong that the fundamental hypotheses of the theory are hardly less certain than observed facts. The probability of the truth of the hypotheses, however, cannot be greater than the improbability that another set of equally simple hypotheses should be conceivable, which should equally well explain all the phenomena. When the hypotheses are of a general and simple character, the improbability in question may become extremely strong; but it diminishes in proportion as the hypotheses become more particular. In sifting the evidence for the truth of any set of hypotheses, it becomes of great importance to consider whether the phenomena explained, or some of them, are explicable on more simple and general hypotheses, or whether they appear absolutely to require the more particular restrictions adopted.[47]

That is, the more particular the hypothesis, the greater the probability that another particular hypothesis will explain the phenomena equally well. Hence, because the eventual possibility is knowledge of 'the real processes of nature', great caution should accompany commitment to highly particular theories.

Though universal in intent, in the 1849 paper the declaration addressed a problem concerning the reflection and refraction of light. For example, Arago had found experimentally that the same proportion of a ray of light was reflected from either side of a surface separating two optical media, and supporters of the undulatory theory had had to assume that one half of an undulation was 'lost' when light was reflected. Stokes showed how both followed from his *general* 'principle of reversion' without the need for *particular* theories about microscopic reality at the surface between two media.

In figure 5.1, the incident ray IA equals $\sin X$, where X equals $\lambda^{-1}(2\pi)(vt-x)$. Let b equal the proportion of a ray reflected from the upper surface in the figure, and let c be the proportion refracted into the lower medium. Let e be the proportion of a ray reflected from the

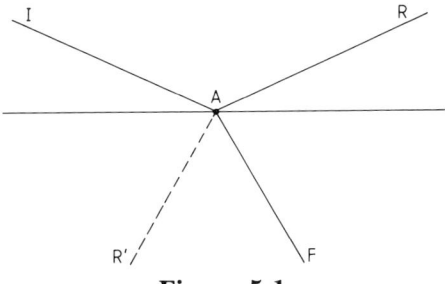

Figure 5.1

lower surface and f the proportion refracted from the lower into the upper medium. Hence, $AR = b \sin X$ and $AF = c \sin X$. Now invoke Stokes's principle of reversion: 'If any material system, in which the forces acting depend only on the positions of the particles, be in motion, if at any instant the velocities of the particles be reversed, the previous motion will be repeated in a reverse order'.[48] Thus, if one imagines AR and AF to be reversed, then they would combine to produce a ray AI, equal in magnitude to IA. If we take into account the various reflections and refractions involved in the process, this simple thought experiment gives:

$$bc \sin X + ce \sin X = AR' = 0 \qquad b = -e \qquad (1)$$

$$b^2 \sin X + cf \sin X = AI = \sin X \qquad cf = 1 - b^2 \qquad (2)$$

'Equation (1) contains at the same time M. Arago's law and the loss of half an undulation'.[49] Equation (2) helped explain the central dark spot in the optical phenomenon known as Newton's rings. The expressions had been accepted before, but were now seen to follow from a general dynamical theory independent of particular theories about dynamical activity at the surface separating the two media. The method provided a surer basis than Fresnel's and Airy's derivation of the expressions which relied on the analogy of colliding elastic bodies of different sizes. However, Stokes did not simply dismiss the Fresnel–Airy approach. As he argued regarding Fresnel's formulae for the intensities of reflected and refracted light, reliance on 'dynamical considerations and analogies' undergirded the resultant formulae by providing a 'dynamical foundation' valid for dynamically conceivable circumstances, 'though whether those circumstances agree with the actual state of reflecting transparent media is another question'.[50] That is, though the Fresnel–Airy approach did provide real support for the formulae, it was wise to have them based on the most *general* principles possible while the search for *particular* theories proceeded. Though one had to be cautious in attributing too much too soon to hypotheses of unobserved reality, nevertheless, given sufficient evidence and insight, one could obtain knowledge of 'the real processes of nature in producing observed phenomena'.

Indeed, fluorescence made it possible to discuss some of nature's real processes. Fluorescence, in its association with 'the inmost structure of chemical molecules', exceeded all other optical phenomena.[51] The undulatory theory supposed light to emanate from 'vibratory movements among the ultimate molecules of the self-luminous body'. It therefore seemed 'natural' to suppose that incident light caused 'vibratory movements' in the molecules of fluorescent substances and that those molecules, 'swinging on their own account', produced ethereal undulations in their turn.[52] Noting that light was even able to produce chemical changes and that most fluorescent substances were complex organic compounds, Stokes reasoned that fluorescence resulted from 'vibrations among the constituent parts of the molecules themselves, performed by virtue of the internal forces which hold the parts of the molecules together'.[53] These vibrations, too small to disrupt the molecule chemically, were sufficient to yield fluorescent light. As James has explained, Stokes conjectured further, seeking the dynamical interaction between ether and matter which would explain his experimental law that in fluorescence light's refrangibility was always lowered.[54] Here, too, Stokes envisaged a dynamical analogy, one which, he told Herschel in 1856, he had used frequently to illustrate his theory. In the analogy a series of waves encounters a row of ships, which not only 'act as a sort of breakwater' but once 'having been sent swinging', themselves 'become centres of disturbance'.

> Now the incident series of waves on the ocean I conceive to be analogous to the incident rays exciting fluorescence, the ships to the material particles, their calming effect to the absorption of the exciting rays which always takes place when fluorescence is produced, the waves emanating from the ships to the fluorescent light.[55]

Others of Stokes's researches could also be placed within the framework of his cautious realism. As to caution, there was his work on the figure of the earth: 'I succeeded in proving Clairaut's Theorem without any hypothesis as to the original fluidity of the Earth, or as to the constitution of its interior';[56] and his improvement on Duhamel's work on the conduction of heat in crystals: 'Duhamel according to the French fashion works on the hypothesis of intra molecular radiation, which for my own part I don't much like making the results depend upon.'[57] Then there was Stokes's research on the ether and stellar aberration—the topic of Chapter 6—where his cautious realism defined a fundamental threat to the undulatory theory of light and then guided his response to it.

Thomson

Thomson's early research, in contrast to Stokes's, has attracted a good deal of recent scholarship, much of which has concerned the role of unobservable entities in his work on heat, electricity and magnetism.[58] Very much influenced by Fourier, Thomson during the 1840s sought 'mathematical theories' of these phenomena, usually avoiding 'hypotheses' about the nature of heat, electricity and magnetism. Not a Comtean-like positivist, Thomson generally cited the lack of sufficient evidence 'in the present state of science' as justification for avoiding hypotheses. Though extremely cautious in committing himself to particular theories of the unobservable, preferring to establish mathematical theories whose validity was independent of such considerations, Thomson was a realist in that he thought such knowledge was possible. With better evidence, therefore, most noticeably in regard to heat at the end of the decade, Thomson began thinking more systematically about nature's hidden realm and went on to devote much speculative effort to such issues for decades thereafter. Knudsen, for example, argues that through most of his career, 'Thomson operated with a hierarchy of three types of theoretical construction.' They were (1) mathematical analogy, where one utilises mathematical similarities between different fields to help fashion mathematical theories, (2) physical analogy or mechanical representation, where one defines a mechanical system analogous to physical phenomena, and (3) dynamical theory, where one identifies the actual, dynamical structure of the unobservable physical world. Thomson thought that the third level, though difficult to reach, was obtained on occasion, for instance, as Knudsen says, in the case of 'the nature of magnetism and its interaction with light'.[59] Concentrating on the third level, this section discusses Thomson's early opinions about unobservables in heat, electricity and magnetism, relating them to his remarks on the luminiferous ether.

He rejected theories of magnetic fluids. In 1845, discussing magnetic 'images' by analogy with optical images, Thomson wrote of 'a mass of positive or negative magnetism', by which, he pointed out, he did not mean to 'imply any hypothesis of a magnetic matter or of a fluid or fluids'. It was simply a convenient and brief way of referring to previous findings that the action of a magnetised body 'may always be

Stokes in 1857 at the age of 38. From J Larmor (ed) 1907 *Memoir and Scientific Correspondence of the Late Sir George Gabriel Stokes*, 2 vols (Cambridge: Cambridge University Press) vol. I frontispiece.

Thomson in 1859 at the age of 35. Photograph by the Rev. David King, Thomson's brother in law. Thomson is reading a letter from Fleeming Jenkin. From Agnes Gardner King 1925 *Kelvin the Man* (London: Hodder and Stoughton) frontispiece.

represented by an imaginary positive and negative distribution of matter, of which the whole mass is algebraically nothing'.[60] In his 'Mathematical Theory of Magnetism' of 1849, he declared that he wanted to establish his theory independently of a 'hypothesis of two magnetic fluids' which previous mathematical theories had 'adopted, and strictly adhered to throughout'. Thomson thought that 'no physical evidence can be adduced in support of such a hypothesis; but on the contrary, recent discoveries, especially in electro-magnetism, render it extremely improbable'.[61] He wrote to John Tyndall that he was convinced of the 'falseness' of the hypothesis.[62]

Thomson wrote similarly concerning the mode of transmission of electrical action in his discussion in 1845 of Faraday's and Coulomb's ideas. Though Coulomb's laws of electrical force 'naturally suggest the idea of material particles attracting or repelling one another at a distance', Coulomb had wisely avoided that hypothesis.[63] Faraday argued that his work showing that dielectrics influenced the strength of electrical forces established that electrical action was transmitted through series of contiguous particles, not at a distance. As with Coulomb's findings, Thomson maintained that Faraday's only *suggested* action between contiguous particles but did not require that concept. Faraday's and Coulomb's respective researches stood for two equivalent ways of viewing the phenomena, either of which could serve as the basis of a mathematical theory. For Thomson in 1845, the mathematical theory of electricity thus existed independently of any particular hypothesis about the nature of electrical action. However, Thomson thought that such hypotheses might be discovered in the future. It was no doubt possible, for example, that electrical forces, seeming to act at a distance,

> may be discovered to be produced entirely by the action of contiguous particles of some intervening medium, and we have an analogy for this in the case of heat, where certain effects which follow the same laws are undoubtedly propagated from particle to particle. It might also be found that magnetic forces are propagated by means of a second medium, and the force of gravitation by means of a third. We know nothing, however, of the molecular action by which such effects could be produced, and in the present state of physical science it is necessary to admit the known facts in each theory as the foundation of the ultimate laws of action at a distance.[64]

Given science's continued limitations, Thomson published 'A Statement of the Principles on Which the Mathematical Theory of Electricity is Founded' in 1848 because, he said, there existed no other 'published work in which the principles are stated in a sufficiently concise and correct form, independently of any hypothesis, to be altogether satisfactory in the present state of science'.[65]

Thomson's thoughts on the nature of heat passed through three stages. First, in the context of Carnot's theory of heat, he thought that heat was conserved. It is less clear whether he also followed Clapeyron and Carnot in regarding heat as a special substance, caloric. Consideration of Joule's work presented at the 1847 meeting of the British Association brought on the second stage, one of growing indecision about the conservation of heat and increasing reflection on the nature of heat. Third, as presented in his great paper of 1851, came his reconciliation of the best of Carnot with Joule's results, through his abandonment of the conservation of heat and his acceptance of Joule's dynamical theory of heat.

Before Joule's arguments had their effect, did Thomson accept the caloric theory? If he did, he would have been following Thomas Thomson, to the extent that Thomas, though undecided, at least viewed the theory of heat as an imponderable substance as a legitimate alternative. If he did not, he would have been following Meikleham's more sceptical view, doubtful that imponderable *substances* even existed. Though some of Thomson's later statements (discussed below) make it possible that he accepted the caloric theory during this earlier period, perhaps the best insight into Thomson's view in the mid-1840s is contained in two statements recorded by one of his students in the autumn of 1847, while he was still resisting Joule's conclusions:

> The standard quantity of Caloric (or heat) is the quantity of heat required to melt a pound of ice. *Heat* and *Caloric* are synonymous terms.

> We will not ask with Dr Black whether the heat is a distinct body, or whether it has a repulsive force, or the power of weakening the Cohesion.[66]

His reluctance to ask whether heat were a distinct body indicates that he was using *caloric* synonymously with *heat* to mean measurable heat, not an imponderable substance.[67] Thus, Thomson could refer not only to electricity and magnetism, but to caloric (i.e. heat) as well, without necessarily implying the existence of imponderable substances.

In 1849, however, Thomson asked the question as Thomas Thomson had: was heat a substance or a quality? Noting the importance of Joule's work, Thomson argued that

> the necessity of a most careful examination of the entire experimental basis of the theory of heat has become more and more urgent. Especially all those assumptions depending on the idea that heat is a *substance*,

invariable in quantity; not convertible into any other element, and incapable of being *generated* by any physical agency; in fact the acknowledged principles of latent heat; would require to be tested by a most searching investigation before they ought to be admitted, as they usually have been, by almost every one who has been engaged on the subject, whether in combining the results of experimental research, or in general theoretical investigations.[68]

His statement that 'almost every one' accepted the idea of heat as a substance may indicate that he had, in fact, supported this view himself earlier. At the least, the quotation shows how Joule's findings were forcing Thomson to consider the issue of the nature of heat more explicitly than before. The 1849 paper assumed that Carnot was right. Assuming in 1851 that Joule was right, Thomson posed the issue as one of heat as substance *versus* heat as motion. He considered it established 'that heat is not a substance, but a dynamical form of mechanical effect'—'a motion excited among the particles of bodies'.[69]

As in electricity and magnetism, so also with heat, it seems the issue was not the philosophical impossibility of knowledge of the unobservable, but a matter of available evidence. When Joule's arguments, and evidence, made it easier, and more important, to discuss the nature of heat, Thomson did so. Even in his 1851 paper, however, his primary aim was not the explication of the situation at the unobservable level itself. The two propositions of thermodynamics may have required heat to be motion, but not that it be any specific kind of motion.[70]

In this regard, Thomson's comments on the ether, though brief, actually constituted a more ambitious attempt to get at a specific portrayal of nature's hidden realm. Not doing research in optics, Thomson had less occasion than Stokes in the 1840s to publish on the fundamentals of that topic. However, by the mid-1840s, Thomson had heard Nichol's professorial lectures praising the undulatory theory, had been immersed in the physical optics of the Cambridge curriculum, and had collaborated with Stokes, champion of the undulatory theory and, as the next chapter discusses, proponent of a realistic view of the ether. Stokes's major 1845 paper on fluid friction concluded by attributing properties of an elastic solid to the ether.[71] Considering this background plus the evidence existing in favour of the undulatory theory by the mid-1840s, it would be surprising if Thomson did not accept the theory by then, even though, so far as I know, no explicit statement by him to that effect survives. In the autumn of 1846, inspired by Faraday's discovery of the rotational action of magnetism on polarised light, Thomson wrote his own paper on an elastic-solid medium, making use of Stokes's earlier paper and labelling the medium's various strains a 'mechanical representation' of electrical and magnetic phenomena.[72] Evidencing the common context of their work, Stokes wrote to Thomson about the latter's paper: 'Perhaps the jelly-like fluid that we once spoke of may be made in your hands to explain the law of the mutual action of electric currents and the phenomena on the induction of these currents'.[73] Sending a copy of his paper to Faraday, Thomson noted its highly provisional quality, but also raised the possibility of an eventual physical theory that would embrace the undulatory theory of light. 'If such a theory [of the propagation of electric and magnetic forces] could be discovered', he wrote to Faraday, 'it would also, when taken in connection with the undulatory theory of light, in all probability explain the effect of magnetism on polarized light.'[74] In similar vein, Thomson wrote two years later to Stokes: 'When I wrote the paper I had some hope, w^h I still retain, that a satisfactory physical theory of all those agencies, including besides light, is approachable.'[75]

It would seem, then, that in the mid-1840s both Stokes and Thomson were writing realistically about an elastic-solid ether. However, for Stokes, dealing with light, the realism was more pronounced. For Thomson, dealing mainly with less well understood branches of physics, electricity and magnetism, the realism was much more cautious. Still, Thomson here offered the possibility of a more detailed picture of unobservable nature than in his longer works on mathematical theories of electricity and magnetism, or even in his various comments on the nature of heat as late as 1851.

Thomson also linked light and radiant heat, in his lectures as early as 1847 and in his paper 'On the Dynamical Theory of Heat' in 1851. In 1847, he told his class that, because of the similar properties of light and radiant heat, 'whatever *Cause* is attributed to light the same must be attributed to heat'.[76] In 1849, he declared that 'radiated heat follows the same laws as the propagation of light' and that 'we must consider the laws of radiant heat along with Optics'. In discussing the definition of matter in the same set of lectures, he explained how the concept of matter was involved in the particle and undulatory theories of light. On the undulatory theory: 'It is absolutely necessary to suppose a medium. We must suppose a kind of matter which is affected only by light.'[77] Such considerations underlay his published statement in 1851, which related the phenomenon of polarisation (instrumental in establishing the modern undulatory theory of light) to radiant heat:

> The dynamical theory of heat . . . is extended to radiant heat by the discovery of phenomena, especially those of the polarization of radiant heat, which render it excessively probable that heat propagated through "vacant space", or through diathermanic substances, consists of waves of transverse vibrations in an all-pervading medium.[78]

Or, as he echoed his 'mechanical representation' paper to his class in 1852, 'an incompressible elastic solid' was the kind of medium required for the vibrations of light and heat.[79]

In summary, Thomson's realism in the 1840s looked even more cautious than Stokes's because he was exploring areas less well understood 'in the present state of science'. Neither a Comtean-like positivist nor an extreme commonsense sceptic, he regarded the question as one of which inferences were supported by the available evidence. Hence, little could be said concerning the nature of electricity and magnetism. Joule's evidence and arguments eventually led to Thomson's change of mind about the nature of heat. Faraday's experiment linking magnetism and polarised light allowed an approach through electricity and magnetism to the ether, where Thomson's provisional 'mechanical representation' raised hopes of an eventual, detailed physical theory. Whatever the exact influences on their respective views, and however their influence on one another may have fostered similarity of view, Thomson's cautious realism was, it seems to me, much the same as Stokes's, manifesting itself differently, especially with respect to the ether, because Thomson did research in different areas of physics than Stokes.

The theological context

Stokes and Thomson did not write about direct connections between what I have called their cautious realism and broader philosophical and theological considerations. Hence, any attempt to define a philosophical or theological justification for cautious realism must be somewhat problematic. However, given their commitment especially to theological positions, one is obliged to pursue the point. At the least, one can establish their acceptance of a natural theology of mind, which *accompanied* their acceptance of a cautious realism. The further question is whether there was any connection between these two views. The matter of realism is not, of course, a straightforward scientific question to be answered in the technical terms of physics. This section proposes a theological answer, one rooted in theological views of the early-Victorian period.

One component of the early-Victorian design argument, one no doubt reinforced by the Christian idea that man was created in God's image, was that man's mind was designed by God, even that it was specifically designed to be capable of obtaining knowledge of nature. The view was current in both Glasgow and Cambridge during the

1830s and 1840s. The Glasgow professor of moral philosophy told Thomson's class: 'Mind is as full of marks of design as matter. We will find the same adaptations of means to an end, as we see in the material world.'[80] Professor Nichol wrote that man used the mental abilities given him by God in order to investigate nature, and investigating nature was, ultimately, a theological enterprise because it informed man about God. Writing of the vastness of the universe and the variety it displayed, he declared: 'How worthy of the Being on whom the CREATOR has bestowed the great gift of Reason, to seek humbly to trace his footsteps through this maze.'[81] Cambridge men expressed similar views. According to Whewell:

> the creator of the atmosphere and of the material universe is the creator of the human mind, and the author of those wonderful powers of think-ing, judging, inferring, discovering, by which we are able to reason concerning the world in which we are placed; and which aid us in lifting our thoughts to the source of our being himself.[82]

Reviewing one of Whewell's works, Herschel stated that both he and Whewell thought that 'the mind of man is represented as in harmony with universal nature; that we are consequently capable of attaining to real knowledge; and that the design and intelligence which we trace throughout creation is no visionary conception, but a truth as certain as the existence of that creation itself'.[83] Mary Somerville, friend of Herschel, put it this way: 'This mighty instrument [mathematics] of human power itself originates in the primitive constitution of the human mind, and rests upon a few fundamental axioms, which have eternally existed in Him who implanted them in the breast of man when he created him after His own image.'[84] In short, the Creator of mind and nature had designed them for each other.

Thomson and Stokes argued similarly. In the introductory lecture which Thomson delivered each year for a half century to his natural philosophy class, he declared:

> Whether in the active investigations by which we arrive at truth, or in the gratification felt in the possession of knowledge, the intellectual value of science is conspicuous, and the adaptation of the human mind for such enjoyment is a manifestation not less remarkable of the divine goodness than the various arrangements by which the physical powers of the animal world are produced and maintained.[85]

Again, in support of his side in the geological debates of the 1860s, Thomson proclaimed: 'Creative power has created in our minds a wish to investigate and a capacity for investigating.'[86] Stokes made the point in claiming that scientific knowledge should not conflict with knowledge derived from either revelation or from moral considerations:

> Now truth is one, and cannot contradict itself; nor, again, can we imagine that God would deceive His creatures by giving them faculties the right exercise of which would lead to a conclusion at variance with what He might think fit to communicate to them in some more direct manner, or opposed to something at which they might legitimately arrive by the exercise of faculties of an altogether different character; I refer to the moral faculties.[87]

Moreover, following Whewell, Stokes felt that God himself was responsible for these moral faculties:

> To me it seems to be the simplest to suppose that man's mental powers, as well as his bodily frame, were designed to be what they are. How that design was carried out we have no means of knowing, and it does not concern us to inquire; but, assuming that it was so, I see no difficulty in supposing that man's innate sense of right and wrong was as much impressed upon him, as little the creation of his own will, as his bodily frame. If that be so, we may even look on this innate sense of right and wrong as the will of God written upon the heart.[88]

Stokes and Thomson even thought not only that God had designed man's mind, but that He could intervene directly in man's thinking. In correspondence with his fiancée in 1857, Stokes tried to explain his ideas about love and duty. 'I was capable', he wrote, 'of being moved, mathematically as it were, by the belief that a particular course was right; and I do believe that God put these views in my mind, working by means of that which was in me to supply that which was wanting.'[89] Seeking further to allay the uneasiness that his views evidently caused his fiancée, Stokes later wrote: 'But for my part the belief has never deserted me that God Himself guided me in the matter, and if you can feel the same then trust quietly in Him.'[90] Thomson, in a long footnote to a draft of his paper 'On the Dynamical Theory of Heat', discussed dreams and miracles: 'Although all dreams are sent by God, yet they are not sent to give us any knowledge except of our own minds (and much knowledge of this kind they sometimes give). Any single case which should give us other knowledge than this, would be a miracle.'[91] Such statements do not indicate that Stokes or Thomson thought, for example, that scientific knowledge came directly from God. What they do help to demonstrate is Stokes's and Thomson's ready acceptance of a theology of human thinking, the most conspicuous part of which was a belief that God designed man's mind. That, in itself, is an important part of their world view for us to recognise. But can we go further and relate theological design to scientific realism?

In this regard, I would suggest that there are two important features of Stokes's and Thomson's belief in design. First, the design argument for God's existence and attributes was itself, after all, an induction from the observable to the unobservable, in fact to the greatest Unobservable Entity of all. Moreover, it was an induction whose accuracy was confirmed by scripture, not a point to be lost on Thomson or Stokes. Second, God's design of man's mind for the purpose of understanding nature would seem almost to guarantee that well established theories would truly represent a real world. As Stokes observed, God would not deceive his creatures. For empiricists like Stokes and Thomson, the concept of a non-deceiving God would never warrant a Cartesian *a priori* approach to nature. But it would, I suggest, support an approach which was based on and extended the authenticity of everyday experience. Distant objects, invisible or barely perceptible, become recognisable on closer viewing. Microscopes and telescopes render the unobservable observable. One can predict, as Adams did in the 1840s, the existence of an unobserved planet by studying its effects on observed planets, and then find it with a telescope. Formalised in Newton's third rule of reasoning, this everyday empiricism held that unobservables can be known and that they are essentially like observables. Hence, both as an example of inference to the unobservable and as support for an everyday empiricism, the design argument constituted theological support for Stokes's and Thomson's realism. In turn, the very possibility of such realistic knowledge demanded, as Stokes declared in his paper of 1849, caution in committing oneself to particular theories of the unobservable to the exclusion of others.

Whatever the exact relationship between natural theology and cautious realism, however, that realism provides the framework for discussion in the next two chapters on Stokes's and Thomson's efforts to comprehend the material side of the universe. Chapter 6 concentrates on the beginning and end of Stokes's theory of a jelly ether. Chapter 7 investigates Thomson's long search for the underlying causes of the ether's properties. It concludes by comparing Stokes's and Thomson's respective variations on the theme of cautious realism.

Finally, seeing the main thrust of cautious realism in Stokes's and Thomson's deliberations also helps us recognise two later-century exceptions—violations of the analogy of nature. In Chapter 6, we find Kelvin postulating that two material substances (ether and matter) could occupy the same space at the same time, and in Chapter 7 we discover Stokes questioning how relevant an understanding of ordinary matter was to comprehending the 'mysterious' ether.

1   See, for example, P M Harman 1985 'Edinburgh Philosophy and Cambridge Physics: The Natural Philosophy of James Clerk Maxwell' in Harman (ed) *Wranglers and Physicists: Studies on Cambridge Mathematical Physics in the Nineteenth Century* (Manchester: Manchester University Press) pp. 202–24.

2   *Sir Isaac Newton's Mathematical Principles of Natural Philosophy and His System of the World* translated into English by Andrew Motte in 1729, the translations revised and supplied with an historical and explanatory appendix by Florian Cajori (Berkeley: University of California Press, 1947) pp. 398–400, 547.

3   See D B Wilson 1974 'Kelvin's Scientific Realism: The Theological Context' *The Philosophical Journal* **11** 42–3.

4   The discussion draws especially on R Olson 1975 *Scottish Philosophy and British Physics, 1750–1880: A Study in the Foundations of the Victorian Scientific Style* (Princeton: Princeton University Press). See also L L Laudan 1970 'Thomas Reid and the Newtonian Turn of British Methodological Thought' in R E Butts and J W Davis (eds) *The Methodological Heritage of Newton* (Oxford: Blackwell) pp. 103–31, and P M Heimann and J E McGuire 1971 'Newtonian Forces and Lockian Powers: Concepts of Matter in Eighteenth-Century Thought' *Hist. Stud. Phys. Sci.* **3** 265–8.

5   J W Herivel 1966 'Aspects of French Theoretical Physics in the Nineteenth Century' *Br. J. Hist. Sci.* **3** 109–32.

6   On French physical thought see R Fox 1974 'The Rise and Fall of Laplacian Physics' *Hist. Stud. Phys. Sci.* **4** 89–136, and R H Silliman 1974 'Fresnel and the Emergence of Physics as a Discipline' *Hist. Stud. Phys. Sci.* **4** 137–62.

7   T Thomson 1840 *An Outline of the Sciences of Heat and Electricity* 2nd edn (London: H Ballière) pp. 320–1.

8   *Ibid.* p. 281.

9   *Ibid.* p. 286.

10   *Ibid.* p. 284.

11   *Ibid.* p. 288.

12   R Pollok 'Notes Taken in the Natural Philosophy Class of the University of Glasgow—W Meikleham Professor' 12 November 1821 *GUL* MS Gen. 1355/104, and Pollok 'Of Inertia' 22 November 1821 *GUL* MS Gen. 1355/60.

13   W Thomson, Notes on the natural philosophy class, 1839–40, 5 November 1839 *CUL Kelvin Collection* NB9.

14   W Sommerville 'Exercises in Natural Philosophy Prescribed by Dr Meikleham Prof. of Nat. Phil. in the University of Glasgow during the Session of 1818–19': 'Of Friction' and 'On the Laws of the Propagation of Light' *GUL* MS Gen. 608; Pollok 'Notes Taken in the Natural Philosophy Class' 12 November 1821; and Pollok 'Solidity' 15 November 1821 *GUL* MS Gen. 1355/58.

15   Thomson, Notes on the natural philosophy class (note 13), evening of 5 November 1839.

16 J P Nichol, Corrected text of twenty lectures delivered in Montrose in 1829 *GUL* MS Gen. 1570.

17 G B Airy 1842 *Mathematical Tracts on the Lunar and Planetary Theories, the Figure of the Earth, Precession and Nutation, the Calculus of Variations, and the Undulatory Theory of Optics* 3rd edn (Cambridge: J J Deighton) pp. v–vi.

18 *Ibid.* §§ 10–12, 28, 31, 36, 38, 103–9, 128.

19 Stokes, Notes on Hopkins's lectures 'Sound and Light I' *CUL Stokes Collection* PA20.

20 J Z Buchwald 1980 'Optics and the Theory of the Punctiform Ether' *Arch. Hist. Exact Sci.* **21** 245–78.

21 G Green 1871 'On the Laws of the Reflection and Refraction of Light at the Common Surface of Two Non-Crystallized Media' in N M Ferrers (ed) *Mathematical Papers of the Late George Green* (London: Macmillan) p. 245.

22 J Challis 1838 *Syllabus of a Course of Experimental Lectures on the Equilibrium and Motion of Fluids, and on Optics* (Cambridge: J and J J Deighton) p. 22.

23 Stokes 'Prof. Challis's Lectures, Notes' *CUL Stokes Collection* PA35.

24 On Whewell see R E Butts 1965 'Necessary Truth in Whewell's Theory of Science' *Am. Phil. Q.* **2** 1–21 and D B Wilson 1974 'Herschel and Whewell's Version of Newtonianism' *J. Hist. Ideas* **35** 79–97.

25 W Whewell 1834 *Astronomy and General Physics Considered with Reference to Natural Theology* 3rd edn (London: William Pickering) p. 141.

26 D Brewster 1842 Review of Whewell *The Philosophy of the Inductive Sciences, Founded upon Their History, Edinburgh Review* **74** 160. See G N Cantor 1975 'The Reception of the Wave Theory of Light in Britain: A Case Study Illustrating the Role of Methodology in Scientific Debate' *Hist. Stud. Phys. Sci.* **6** 122–3 and G N Cantor 1983 *Optics after Newton: Theories of Light in Britain and Ireland, 1704–1840* (Manchester: Manchester University Press, pp. 180–1. Cantor discusses Brewster's debt to Scottish commonsense philosophy.

27 W F Cannon 1961 'John Herschel and the Idea of Science' *J. Hist. Ideas* **22** 215–39.

28 *Thompson* I 250.

29 Herschel 1966 *A Preliminary Discourse on the Study of Natural Philosophy* (New York: Johnson Reprint Corp.) §§202, 204, 208.

30 *Ibid.* §216.

31 See Wilson 'Herschel and Whewell's Newtonianism' (note 24) pp. 90–4.

32 Herschel *Preliminary Discourse* §142.

33 Brown's influence on Herschel is discussed in Olson *Scottish Philosophy and British Physics* (note 4) pp. 252–70.

34 Stokes 'Theory of the friction of fluids in motion' *CUL Stokes Collection* PA224.

35 Stokes 'On Some Cases of Fluid Motion' *MPP* I 18.

36 *Ibid.* p. 53.

37 Stokes 'On the Theories of the Internal Friction of Fluids in Motion, and of the Equilibrium and Motion of Elastic Solids' *MPP* I 79.

38   *Ibid.* p. 78.
39   *Ibid.* pp. 84–6. In his unpublished manuscript, he stated: 'To recapitulate: the following are the very simple hypotheses on which it now appears that the equations of motion which have been found rest. That a fluid is composed of molecules, perhaps separated from each other, perhaps, in some cases at least, in contact, but smooth if the latter be the case: that when the average of a great number is taken there is no particular arrangement of the molecules in one direction rather than in another: and that if a single molecule were displaced from its position of relative equilibrium, the accelerating force of restitution would be very large unless the displacement were very small.' (Stokes 'Theory of the friction of fluids in motion' (note 34).)
40   Stokes 'On the Theories of the Internal Friction of Fluids' pp. 78–9.
41   *Ibid.* p. 75.
42   Stokes 'Theory of the friction of fluids in motion' (note 34).
43   Stokes 'On the Theories of the Internal Friction of Fluids' pp. 84–6. I have tried to relate Stokes's work in hydrodynamics to the history of atomic theories in D B Wilson 1981 'Kinetic Atom' *Am. J. Phys.* **49** 217–22.
44   Stokes 'Report on Recent Researches in Hydrodynamics' *MPP* I 182.
45   *Ibid.* p. 184.
46   Stokes to Thomson, 10 and 11 March 1888 *CUL Kelvin Collection* S481.
47   Stokes 'On the Perfect Blackness of the Central Spot in Newton's Rings, and on the Verification of Fresnel's Formulae for the Intensities of Reflected and Refracted Rays' *MPP* II 96–7.
48   *Ibid.* p. 90.
49   *Ibid.* p. 92.
50   *Ibid.* p. 98.
51   Stokes 'On the Change of Refrangibility of Light' *MPP* III 390.
52   *Ibid.* p. 388.
53   *Ibid.* pp. 389–90.
54   F A J L James 1983 'The Conservation of Energy, Theories of Absorption and Resonating Molecules, 1851–1854: G. G. Stokes, A. J. Ångström and W. Thomson' *Not. Rec. R. Soc.* **38** 79–107. The article deals mainly with the development of Stokes's ideas, showing, for example, how he took the conservation of energy into account.
55   Stokes to J F W Herschel, 27 June 1856, *Royal Society of London Herschel Papers* H.S. 17.31. Also quoted in James 'The Conservation of Energy' p. 90.
56   Stokes to Thomson, 29 March 1849 *CUL Kelvin Collection* S346.
57   Stokes to Thomson, 6 January 1851 *CUL Kelvin Collection* S360.
58   J Z Buchwald 1977 'William Thomson and the Mathematisation of Faraday's Electrostatics' *Hist. Stud. Phys. Sci.* **8** 102–36; D Gooding 1980 'Faraday, Thomson and the Concept of the Magnetic Field' *Br. J. Hist. Sci.* **13** 91–120; O Knudsen 1976 'The Faraday Effect and Physical Theory, 1845–1873' *Arch. Hist. Exact Sci.* **15** 235–81; O Knudsen 'Mathematics and Physical Reality in William Thomson's Electromagnetic Theory' in Harman (ed) *Wranglers and Physicists* (note 1)

pp. 149–79; D F Moyer 1978 'Continuum Mechanics and Field Theory: Thomson and Maxwell' *Stud. Hist. Phil. Sci.* **9** 35–50; D M Siegel 1981 'Thomson, Maxwell and the Universal Ether in Victorian Physics' in G N Cantor and M J S Hodge (eds) *Conceptions of Ether: Studies in the History of Ether Theories 1740–1900* (Cambridge: Cambridge University Press) pp. 239–68; C W Smith 1976 'William Thomson and the Creation of Thermodynamics' *Arch. Hist. Exact Sci.* **16** 231–88; C W Smith 1978 'A New Chart for British Natural Philosophy: The Development of Energy Physics in the Nineteenth Century' *Hist. Sci.* **16** 231–79; M N Wise 1979 'William Thomson's Mathematical Route to Energy Conservation: A Case Study of the Role of Mathematics in Concept Formation' *Hist. Stud. Phys. Sci.* **10** 49–83; and M N Wise 1981 'The Flow Analogy to Electricity and Magnetism, Part 1: William Thomson's Reformulation of Action at a Distance' *Arch. Hist. Exact Sci.* **25** 19–70.

59   Knudsen 'Mathematics and Physical Reality' (note 58) pp. 166–7.
60   Thomson 'Note on Induced Magnetism in a Plate' *PEM* p. 105.
61   Thomson 'A Mathematical Theory of Magnetism' *PEM* p. 344.
62   Thomson to Tyndall, 14 August 1850 *Royal Institution Tyndall Papers*.
63   Thomson 'On the Elementary Laws of Statical Electricity' *PEM* p. 15.
64   *Ibid.* p. 37.
65   Thomson 'A Statement of the Principles on Which the Mathematical Theory of Electricity is Founded' *PEM* p. 42.
66   Anonymous notes on Thomson's lectures for 1847–1848 *GUL* MS Gen. 627 and MS Gen. 633.
67   This may, then, be the same sense in which he used the term *caloric* in 1848 in 'On an Absolute Thermometric Scale Founded on Carnot's Theory of the Motive Power of Heat, and Calculated from Regnault's Observations' *MPP* I 102: 'The conversion of heat (or *caloric*) into mechanical effect is probably impossible, certainly undiscovered.'
68   Thomson 'An Account of Carnot's Theory of the Motive Power of Heat; with Numerical Results Deduced from Regnault's Experiments on Steam' *MPP* I 116.
69   Thomson 'On the Dynamical Theory of Heat, with Numerical Results Deduced from Mr Joule's Equivalent of a Thermal Unit, and M. Regnault's Observations on Steam' *MPP* I 174–5.
70   He did speculate on the question during the 1850s, envisaging heat to be caused by rotational motion of some kind. (Thomson to J P Joule, 31 March 1852, in Joule 1853 'On the OEconomical Production of Mechanical Effect from Chemical Forces' *Phil. Mag.* **5** 3n and Thomson 1856 'Dynamical Illustrations of the Magnetic and Helicoidal Rotatory Effects of Transparent Bodies on Polarized Light' *Proc. R. Soc.* **8** 152.)
71   Stokes 'On the Theories of the Internal Friction of Fluids' (note 37) pp. 124–9.
72   Thomson 'On a Mechanical Representation of Electric, Magnetic, and Galvanic Forces' *MPP* I 76–80.
73   Stokes to Thomson, 13 March 1847 *CUL Kelvin Collection* S323.
74   Thomson to Farady, 11 June 1847, in *Thompson* I 203–4.

75   Thomson to Stokes, 4 February 1849 *CUL Stokes Collection* K28.

76   Notes on Thomson's lectures (note 66).

77   W Smith, Notes on Thomson's lectures for 1849–1850, 2 November 1849 *GUL* MS Gen. 142.

78   Thomson 'On the Dynamical Theory of Heat' (note 69) p. 174.

79   W Jack, Notes on Thomson's lectures for 1852–1853 *GUL* MS Gen. 130.

80   Thomson, Notes on the lectures of William Fleming *CUL Kelvin Collection* NB21.

81   J P Nichol 1838 *The Phenomena and Order of the Solar System* (Edinburgh and London) pp. 144–5.

82   Whewell *Astronomy and General Physics* (note 25) pp. 257–8. On Whewell and Herschel, see Wilson 'Herschel and Whewell's Newtonianism' (note 24) and J L Richards 1979 'The Reception of a Mathematical Theory: Non-Euclidean Geometry in England, 1868–1883' in B Barnes and S Shapin (eds) *Natural Order: Historical Studies of Scientific Culture* (Beverly Hills and London: Sage) pp. 146–50.

83   J F W Herschel 1857 'Whewell on the Inductive Sciences' in Herschel *Essays from the Edinburgh and Quarterly Reviews* (London) p. 246.

84   M Somerville 1836 *On the Connexion of the Physical Sciences* 3rd edn (London: John Murray) p. 412.

85   *Thompson* I 246.

86   Thomson 'On Geological Time' *PLA* II 45. His point was that even though God *could* have made a sun that would shine undiminished forever (in agreement with uniformitarian geology), he gave man the intelligence to work out that he did not do so.

87   Stokes 1893 'Science and Faith' *The Official Report of the Church Congress* 340–1.

88   Stokes *Natural Theology* I 231.

89   Stokes to Mary Susanna Robinson, 21 January 1857 *Memoir* I 52–3.

90   Stokes to Mary Susanna Robinson, 13 June 1857 *Memoir* I 69–70. Stokes had unsettled his fiancée with a 55 page letter (which does not survive) on his views of love and duty, evidently expressing more duty than love towards her.

91   Entry dated 1 March 1851 in Thomson, Draft of 'On the Dynamical Theory of Heat' *CUL Kelvin Collection* PA128.

6

Stokes's Jelly Ether

Writing in 1903 of the establishment of the elastic-solid theory of light, J T Merz declared: 'The natural philosopher to whom we are most indebted for bringing clearness and definiteness into our ideas and our language in these very intricate subjects is Sir George Stokes.' Up to the time of Stokes, Merz thought, 'the ether was universally spoken of as a fluid'.[1]

Though somewhat exaggerated, Merz's account contains enough truth to suggest why Stokes felt in the 1840s that he had to *argue for* the concept of an ether with solid properties. The possibility that the ether might resemble a solid arose when Thomas Young and Augustin Fresnel decided that the phenomena associated with polarised light must be caused by transverse, not longitudinal, waves of light. Among ordinary materials, solids, not fluids, support transverse waves. It may seem logical, therefore, that the idea of transverse waves would lead naturally to the well known, nineteenth-century programme of research on an elastic-solid ether.[2] Historically, however, the concept met resistance. Young himself, for example, found the concept 'perfectly *appalling*' and argued that 'the hypothesis [that fluids can support transverse vibrations] remains completely open for discussion, notwithstanding the apparent difficulties attending it'.[3] Thus, in his theory of stellar aberration of the 1840s, Stokes was not simply working out details and implications of an already well accepted concept of the ether. Rather, his theory of aberration was part of his attempt to demonstrate the concept's validity. Stokes's resultant theory of a 'jelly' ether was successful enough to help make the elastic-solid theory 'the most celebrated special form of the wave theory'.[4] In later years, the jelly ether's most celebrated proponent was Kelvin.

This chapter concentrates on the origin of Stokes's jelly ether in the 1840s. After description of the French background to Stokes's theory, its institutional and conceptual setting is discussed. From a distinctly Cambridge perspective, Stokes responded to French ideas, formulating an ether theory which, with modification, lasted, in Kelvin's mind, into the twentieth century. The chapter's conclusion looks at these final years of the theory.

The French background

In 1818 Fresnel presented his theory of the relationship between the ether and material bodies moving through it.[5] His theory explained both stellar aberration and phenomena associated with transparent bodies in motion with respect to the ether.

The explanation of aberration adopted by Fresnel had been proposed previously in terms of both the corpuscular and wave theories of light. In publishing his discovery of aberration in 1728, James Bradley said that the effect resulted from the combination of the observer's motion and that of the light particles.[6] That is, when one observes a star, one must tilt one's telescope slightly so that the stellar light can get through the telescope without hitting the side of the telescope. Bradley stated that the tilting is required because the observer is moving with the earth at the same time as particles of stellar light are moving towards the observer. Although similar to the explanation of aberration based on the corpuscular theory, the wave-theory explanation involved an additional, fundamental assumption—that the ether passes unhindered through the earth. If the ether were disturbed by the earth's motion, the propagation of a wave of light would no longer follow the same path as a corpuscle of light. The disturbance in the ether would deflect the light out of its rectilinear path, producing complicated and unpredictable results. Aware of these potential difficulties, Young concluded in 1804: 'Upon considering the phenomena of the aberration of the stars, I am disposed to believe, that the luminiferous ether pervades the substance of all material bodies with little or no resistance, as freely perhaps as the wind passes through a grove of trees.'[7]

Challenging Young's conclusion was an experiment performed in 1810 by François Arago.[8] Arago reasoned that, because of the ether's free passage through the earth, there should be a relative velocity between the ether and the earth at the surface of the earth equal to the velocity of the earth around the sun. Therefore, if one could measure the velocity of light from a star towards which one was moving, one would find it to be greater than the velocity of light from a star away from which one was moving. Light from both stars should travel at the same velocity within the ether, and the velocity measured by the observer on earth should equal the velocity of light within the ether *plus* or *minus* the velocity of the ether with respect to him. Two rays of light moving at different velocities should have been refracted differently. However, Arago's experiments disclosed no such difference—that is, they detected no relative motion between the ether and the earth at the earth's surface.

In his attempt to account for Arago's results, Fresnel posed the problem with which any adequate theory of the ether would have to deal:

> If we admit that our globe imparts its motion to the ether by which it is enveloped, we may easily conceive why the same prism always refracts light in the same manner, from whatever side it arrives. But it appears impossible to explain stellar aberration on this hypothesis. Up to the present at least, I have only been able clearly to conceive this phenomenon by supposing that the ether passes freely through the globe.[9]

He resolved the dilemma by supposing a certain quantity of ether to be permanently attached within transparent media. First, he considered a prism at rest with respect to the ether. This situation enabled him to find the ratio of the density of the ether inside the prism to the density of the ether in space. We can do this, he said,

> because we know that it must be inverse to that of the squares of the velocities of propagation of the waves. If d and d' be the wavelengths of light in the ether surrounding and inside the prism, Δ and Δ' the densities of the two *milieux*, we then have the proportion:
>
> $$d^2 : d'^2 :: \Delta' : \Delta \qquad \text{[10]}$$

Consequently, the difference in the density of the ether inside and outside the prism ($\Delta' - \Delta = \Delta[(d^2 - d'^2)/d'^2]$) 'is the density of the mobile part of the *milieu* within the prism'.[11] Therefore, if a prism is moving through the ether, the ether within it will be composed of two parts (1) the free ether of space with density Δ passing unhindered through the prism and (2) the ether of density $\Delta' - \Delta$ which is bound to the prism and goes wherever the prism goes. The total density of ether in a moving prism is Δ', but, because only the bound part of the total amount moves with the velocity of the prism, the apparent velocity of the total amount will be less than the velocity of the prism. This effect came to be expressed in the equation $v' = v(1 - n^{1/2})$, where v is the velocity of the prism, n the index of refraction of the prism, and v' the velocity at which the total ether in the prism appears to be dragged along by the prism. The quantity $(1 - n^{1/2})$ is known as Fresnel's 'dragging coefficient'.[12] With this theory of the ether, Fresnel was able to explain Arago's experiment and, at the same time, to preserve Young's simple explanation of aberration.

Realism and the ether

Hydrodynamics was the first subject of research which Stokes under-took. He had studied hydrodynamics with William Hopkins and perhaps had heard James Challis lecture on the subject.[13] After taking his degree in 1841, Stokes decided to 'try my hand at original research' and, following the suggestion of Hopkins, turned to hydrodynamics, 'then at rather a low ebb in the general reading' at Cambridge.[14] When Stokes completed his first paper, Hopkins gave a copy of it to Challis for his comments. Sharply disputing Stokes's methods and results, Challis precipitated an extended debate between himself and Stokes on hydrodynamical questions which they pursued through several numbers of the *Philosophical Magazine* and in a correspondence of some 21 letters.

The initial dispute concerned the physical significance for fluid motion of the expression $udx + vdy + wdz$ being an exact differential, where u, v and w were, respectively, the components of the fluid's velocity along the x, y and z coordinate axes. Having called attention to the question in earlier writings, in February of 1842 Challis pub-lished his answer—when $udx + vdy + wdz$ was exact, the motion of the fluid was rectilinear.[15] By contrast, Stokes's first paper, completed in April of 1842, sought to prove that the expression was exact 'in the case of steady motion, when the lines of motion are open curves'.[16] He read the paper to the Cambridge Philosophical Society on 25 April, and, as Challis reported to Stokes: 'Mr. Hopkins placed in my hands a paper written by you in which the question is considered whether curvilinear motion of a fluid is possible when udx + vdy + wdz is an exact differential of a function of three independent variables.'[17] After discussion and correspondence between the two, Challis, referring to the 'unsoundness of your reasoning', wrote to Stokes in August that, given the subject's importance, Stokes should *publish* any further objections he might have to Challis's work.[18] In October, Stokes's criticism appeared in print,[19] and the debate between the two on this and other hydrodynamical issues continued in print and corres-pondence through most of 1843.

In 1845 Stokes published his own interpretation of the physical meaning of $udx + vdy + wdz$ being an exact differential. It meant that if an 'indefinitely small' portion of the fluid were 'suddenly solidified and detached from the rest of the fluid', it would not rotate at all but would 'move with a motion simply of translation'.[20] The absence of rotation meant that there would be no eddies in the fluid, but did not preclude the fluid's flowing along curved lines. Stokes's became the accepted interpretation, being repeated, for example, in an encyclo-paedia article in the 1870s by Maxwell, who used the still-accepted term for this kind of motion, 'irrotational'.[21]

This, the first of Stokes–Challis disputes, sheds light on the Cambridge context of Stokes's jelly ether. Challis, though prominent, was not highly regarded as a mathematical physicist by the likes of Hopkins, Stokes and Thomson. Writing to Stokes in 1852 about Challis's work on sound, for example, Thomson asked: 'Do you not think that Challis's discovery is really "an aerial bore"? . . . I am strongly impressed with the idea of *bore* connected with various discoveries of Challis's both in hydrodynamics and optics.'[22] One can easily imagine that Hopkins steered his senior wrangler of 1841 towards Challis's speciality and then gave the resultant paper to Challis as an attempt to counter Challis's influence on Cambridge hydrodynamical studies. The pattern of Stokes challenging Challis's positions continued in the case of stellar aberration, for in his lectures Challis discussed the nature of the ether and suggested an explanation of aberration. Further, the dispute over hydrodynamical principles helps mark Stokes's growing expertise and prominence. In 1846 the British Association asked him to write a report on hydrodynamics, after Challis declined.[23] In 1848 and 1849 Stokes and Thomson continued the improvement of Cambridge hydrodynamical studies with six 'Notes on Hydrodynamics' designed for the use of students.[24] Finally, the Challis–Stokes hydrodynamical dispute calls attention to the physical significance of $udx + vdy + wdz$ being an exact differential, a key point in Stokes's realistic portrayal of stellar aberration within a jelly ether.

Stokes's hydrodynamical research was closely involved in his rejection of Fresnel's concept of the unhindered passage of the ether through the earth. Particularly significant in this respect was the problem of the oscillation of a ball pendulum in a fluid. In 1842 or 1843 Stokes's future father-in-law, Thomas Romney Robinson, astronomer in charge of the Armagh observatory, told him about a 'remarkable experiment' performed by James South, the results of which were unpublished. Describing South's experiment in 1850, Stokes wrote:

When a pendulum is in motion, one would naturally have supposed that the air near the moving body glided past the surface, or the surface past it, which comes to the same thing if the relative motion only be considered, with a velocity comparable with the absolute velocity of the surface itself. But on attaching a piece of gold leaf to the bottom of a pendulum, so as to stick out in a direction perpendicular to the surface, and then setting the pendulum in motion, Sir James South found that the gold leaf retained its perpendicular position just as if the pendulum had been at rest; and it was not till the gold leaf carried by the pendulum had been removed to some distance from the surface, that it began to lag behind. This experiment shews clearly the existence of a tangential action between the pendulum and the air, and between one layer of air and another.[25]

The results of this experiment had apparently influenced Stokes before 1850. In 1845, for example, he pointed out that the common hydro-dynamical theory of pendulums was imperfect because it ignored the friction between the pendulum bob and the fluid.[26] Commenting again on this problem in his 1846 report to the British Association, he declared that 'when the tangential action of the sphere on the fluid, and the internal friction of the fluid itself are considered, it is clear that one consequence will be, to speak in a general way, that a portion of the fluid will be dragged along with the sphere'.[27] As we shall see, the idea of a sphere dragging a fluid with it closely resembled Stokes's idea of the relationship between the earth and the ether.

The resemblance also depended, in addition to strictly hydro-dynamical notions, on Stokes's strong adherence to the ether as an actually existing unobservable entity which could be understood partially through analogies with everyday material substances. He defended this view, for example, against the Oxford professor of geometry, Baden Powell, who in commenting on the aberration debate seemed to contrast the ether with 'known' causes like terrestrial magnetism and thus to argue that one could well de-emphasise the ether's role in stellar aberration.[28] In a letter to Powell, Stokes objected:

> It is true the existence of an ether cannot be considered as a *known* cause, yet I cannot but regard the astonishing simplicity of the explanations of optical phenomena afforded by the undulatory theory as rendering the existence of an ether probable in a very high degree. And if an ether does exist, the analogy of all bodies moving through fluids with which we are acquainted would lead us to suppose that the earth in its course round the sun pushed aside the ether, rather than that it let it freely pass thro'.[29]

Therefore, Stokes continued, it was 'no unreasonable demand to see whether' an ether–earth interaction would 'produce a deviation' in the luminous vibrations transmitted by the ether.

Moreover, Stokes felt he had to defend the concept of a solid ether. Though both G B Airy in his textbook and Hopkins in his lectures presented the ether as a medium which could support transverse waves, neither dwelt on the ether's solidity. John Herschel regarded the concept of a solid ether as only a temporary device, useful 'till the real truth shall be discovered', and William Whewell did not write of the ether as a solid.[30] Challis explicitly regarded the ether as a fluid and, in his lectures attended by Stokes, sought to explain polarisation without even employing transverse waves.[31] To be sure, George Green had written a couple of papers which treated the ether as an elastic solid, but he died in 1841.[32] Moreover, Green had not addressed the severe problem of conceptualising the motion of ordinary solids through an elastic-solid ether. Thus, in the Cambridge of the 1840s, Stokes apparently found considerable doubt or disbelief as he continued Green's work on an elastic-solid ether.

Stokes's first published statements about the ether supported its elastic-solid character. Having pointed out the similarities of the equations used for elastic solids and for the ether, Stokes observed that 'it is sometimes brought as an objection to the equations of motion of the luminiferous ether, that they are the same as those employed for the motion of solid bodies, and that it seems unnatural to employ the same equations for substances which must be so differently constituted'.[33] He attempted to undermine this objection with an appeal to the 'law of continuity', reasoning that the property of elasticity should exist in all bodies of varying degrees of solidity from the most solid to the most liquid. If we imagine a solid body to become more and more liquid there is no reason for us to assume that, at some point in its transformation, its elasticity will suddenly disappear. In one degree or another the resistance of a body to a transverse displacement of its parts should exist throughout the transformation. The elasticity may become insensible in ordinary cases and may be masked by more conspicuous properties of a body, but it may still be 'that [property] on which the phenomena of light depend'. We must remember, he pointed out, that whereas the length of a sound wave in air may be several feet the maximum length of a wave of light is about 0.0003 of an inch.

> It is easy to imagine that the relative displacement of the particles of ether may be so small as not to reach, nor even come near to the greatest relative displacement which could exist without the molecules of the medium assuming new positions of equilibrium, or, to keep clear of the idea of molecules, without the medium assuming a new arrangement which might be permanent.[34]

In his 1849 paper on diffraction he still felt he had to defend the concept. So completely did the theory of transverse vibrations conduct 'us through a multitude of curious and complicated phenomena, like a thread through a labyrinth' that its claims to truth were 'but little short of those of the theory of universal gravitation'. But, in accepting this theory, Stokes argued,

> we are obliged to suppose the existence of a tangential force in the ether, called into play by the continuous sliding of one layer, or film, of the medium over another. In consequence of the existence of this force, the ether must behave, so far as regards the luminous vibrations, like an elastic solid.[35]

But Stokes was careful to stop short of a specific theory of the underlying cause of the ether's properties. He wrote to Powell that he was dealing 'geometrically' with the kind of ethereal motion for which $udx + vdy + wdz$ was exact. 'I do not here enter into the question of the mechanical laws w^h may be conceived to govern the motion of a substance of w^h we know so little as we do of the ether'.[36] In addition, in recognising the existence of a tangential force in the solid ether, 'we have no occasion to speculate as to the cause of this tangential force, nor to assume either that the ether does, or that it does not, consist of distinct particles'.[37] In fact, the equations of motion for small transverse vibrations in an elastic-solid medium were consistent with the medium's being either continuous or particulate.[38] As with the ordinary fluids of hydrodynamics, whatever ether's ultimate structure, it could be regarded as continuous. Thus, Stokes was careful to distinguish between what was known about the unobservable ether and what was not.

We can easily see the relevance of what Stokes did know to his investigation of stellar aberration. After a few years of thinking through the complex problems of ordinary fluid–solid interactions, Stokes must have been acutely conscious that there should be a similar interaction between the elastic-solid ether and the moving earth. No matter how negligible the elasticity of the ether might be, elastic-solid properties must have contributed to the amount of viscosity Stokes imagined the ether to possess. And the more viscous the ether appeared to be, surely the more certain was the existence of some friction between the earth and the ether. The cautious realist Stokes simply could not conceive of the unimpeded passage of an elastic-solid ether around, much less through, the earth. The truth of the modern undulatory theory of light entailed the reality of an elastic-solid ether, which in turn entailed considerable interaction between earth and ether, and which finally posed potentially serious difficulties for explaining aberration. Hence, Stokes could tell his class in 1864 that stellar aberration was the one area where the corpuscular theory of light, with its simple explanation, had the advantage over the undulatory.[39] Aberration was obviously a crucial phenomenon to explain, in the right way.

Such considerations explain Stokes's very strong language in rejecting Fresnel's theory of aberration as a 'rather startling hypothesis'[40] and a 'violent supposition'.[41] There was also a vast difference between Challis and Stokes. In the lectures that Stokes attended, Challis suggested an explanation of aberration largely the same as that given on the corpuscular theory of light—it resulted from the speed of light and the speed of the observer. Challis did think it 'possible that the ether may be carried by the atmosphere with part of the vel^y of the earth', but, in Challis's theory, that only reduced the effective velocity of the earth in the calculation of the amount of aberration.[42] Ignored were questions of what sort of ethereal motions such an ether–atmosphere interaction might set off and how they might affect incoming starlight. Challis's view of the ether may have seemed more properly realistic than Fresnel's, but it was not realistic enough. Aberration was thus a focus for a great deal. As French mathematical optics was assimilated into the Cambridge setting, there was the need to correct certain French concepts, to straighten out Challis's distortion of Cambridge studies, and to establish the reality of the elastic-solid ether. What was needed was a successful explanation of stellar aberration, giving central importance to a realistic view of the ether, and all that that implied.

Stokes versus Challis

Stokes read his first paper on aberration to the Cambridge Philosophical Society on 12 May 1845,[43] beginning another conflict with Challis. Both read papers on aberration at the British Association meeting in June 1845.[44] As before, Challis and Stokes carried on this dispute both in private correspondence and in communications to the *Philosophical Magazine*. Between 1845 and 1855 they sent 13 papers and notes to the *Philosophical Magazine*, most of which appeared in 1845 and 1846. In the December 1846 issue of the *Philosophical Magazine* Baden Powell noted the importance of the dispute:

> The explanation of the phaenomenon of the aberration of light, as given in most of the established treatises on astronomy, seems for at least a long time past, to have been generally acquiesced in; until at the meeting of the British Association, 1845, the subject was stirred anew by the announcement of the investigations of Prof. Challis and Mr. Stokes, the discussion of which has been continued in so many numbers of the Philosophical Magazine . . . and which seemed to open a new epoch in the history of the question.[45]

Stokes developed his theory in three major papers. The first was that read in 1845 at Cambridge and to the British Association and published in the *Philosophical Magazine*. In it Stokes determined that aberration could be explained if the motion of the ether, due to the disturbance of the earth, was subject to a certain restriction—that $udx + vdy + wdz$ was an exact differential. In two later papers, published in 1846[46] and 1848,[47] he argued that the kind of ether that would behave in accordance with this restriction was one with solid characteristics.

In his 1845 paper, after rejecting Fresnel's 'rather startling hypothesis', Stokes presented his own hypothesis of the effect of the earth on the ether.

> I shall suppose that the earth and planets carry a portion of the aether along with them so that the aether close to their surfaces is at rest relatively to those surfaces, while its velocity alters as we recede from the surface, till, at no great distance, it is at rest in space.[48]

Hence, he established what amounted to 'boundary conditions' for his problem. The velocity of the ether at the surface of the earth was equal to the velocity of the earth through the stationary ether. The ether 'at no great distance' from the surface of the earth was undisturbed by the motion of the earth—its velocity was zero. Stokes went on to illustrate his conception of the passage of stellar light through the ether near the earth and how that passage produced aberration. Figure 6.1 represents the path of a small portion of a wave of light approaching the earth, and figure 6.2 is an orthogonal trajectory to the wave fronts on which *e* is the point on the earth reached by the light. As the wave traverses the region near the earth, its orientation is changed by the moving ether. Furthermore, 'the direction in which a heavenly body is seen is normal to the fronts of the waves which have emanated from it . . .'.[49] Therefore, when the light wave in figure 6.1 reaches the earth,

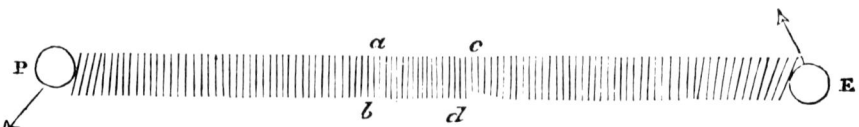

Figure 6.1 From G G Stokes 1845 'On the Aberration of Light' *Phil. Mag.* **27** 13. The figure shows the change in orientation of a small portion of a wave of light travelling from a planet to the earth. '*P* is the planet in the position it had when the light quitted it; *E* the earth in the position it has when the light reaches it.' As Stokes explained in a footnote, the lines near *P* should lean in the opposite direction.

it will appear to have come from the direction perpendicular to the wave fronts. This explains stellar, and planetary, aberration.

To produce this result, the motion of the ether near the earth had to be such that $udx + vdy + wdz$ was an exact differential, when u, v and w were, respectively, the components of the velocity of the ether along the x, y and z coordinate axes. If this were the case, there would be no whirlpools or eddies in the ether as it flowed around the earth. Moreover, this condition provided a necessary link between the size of the angle of aberration and the magnitude of the earth's speed. The situation imagined by Stokes involved an equality determined by the definition of $udx + vdy + wdz$ as an exact differential: $du/dz = dw/dx$. In figure 6.3, AB is a small portion of the earth and PQ is the front of a light wave which is perpendicular to the z-axis and travelling along the z-axis, towards AB. All parts of the wave front are propagated through the ether at the same rate. Then, if the orientation of the front should change to P_1Q_1, for example, the change must be caused by a variable motion in the ether itself. The change of orientation represented by P_1Q_1 would result from differing velocities in the direction of the z-axis of the parts of the ether in PQ—the change in velocity represented by dw/dx. We know that $dw/dx = du/dz$ and, therefore, that w varies along the x-axis just as u does along the z-axis. Furthermore, since u varies from zero to the speed of the earth, the variation of w is related to the earth's speed. Hence, the theory explains why the angle of aberration is determined by the speed of the earth.

Figure 6.2 From G G Stokes 1845 'On the Aberration of Light' *Phil. Mag.* **27** 13. This is the orthogonal trajectory to the lines *ab*, *cd*, etc in figure 6.1. 'The trajectory *pmne* may be considered a straight line, except near the ends *p* and *e*, where it will be a little curved, as from *p* to *m* and from *e* to *n*.' The curve at *pm* had negligible effect on our observation of the planet; that at *ne*, therefore, accounted for essentially all of the aberration. Stokes later realised that *pmne* was only the orthogonal trajectory and did not represent the path of the light, which turned out to be a straight line, not curved as he first thought. He omitted figures 6.1 and 6.2 in the reprint of the article.

Challis's opposition to Stokes turned largely on the question of what constituted an explanation of aberration. Not quite the same as the standard theory of aberration, Challis's explanation took into account

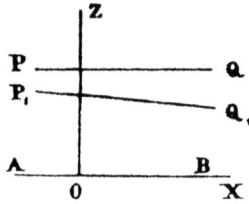

Figure 6.3 From H A Lorentz 'De l'influence du mouvement de la terre sur les phénomènes lumineux' in *Collected Papers* 9 vols (The Hague: Martinus Nijhoff, 1937) IV 159. This is Lorentz's figure to illustrate his discussion of Stokes's theory of aberration.

light coming from both a star and the cross wires of a telescope. It stated that the angular separation between the two objects when their images coincided was equal to the angle of aberration and, further, that the wires, not the star, were seen out of their true position. Employing only the 'known causes' of the speed of light and the speed of the earth, Challis's theory was independent of any ideas about the ether. In fact he thought that such ideas were necessary not to explain aberration but to reconcile the wave theory with his correct explanation of aberration. Replying to a note from Stokes, he wrote:

> I am at a loss to conceive on what grounds you say that "there is nothing in the subject of aberration about which we differ". Our explanations proceed on principles totally at variance. If mine, which refers aberration to known causes be true, then yours, which refers it to hypothetical causes, must be false. My explanation (granting it to be true) makes it impossible that any explanation can be true which requires the consideration in any way whatever of the motion of the aether.[50]

Although Challis apparently did not question the legitimacy of Stokes's arguments, he did not regard them as providing an explanation of aberration. Rather, they were an attempt to reconcile the undulatory theory of light with Challis's explanation. And if such a reconciliation were not possible, 'the undulatory theory, and not the foregoing explanation [of aberration] is at fault'.[51] In subsequent papers, Challis modified the details of his theory and much later noted that the correct version of the theory appeared first in 1852 'after attempts made in 1845 and 1846 with only partial success'.[52] However, Challis's opinion concerning Stokes's theory of aberration remained constant.

Challis's confidence in the undulatory theory assured him that it could be reconciled with his theory of aberration.[53] He agreed with Stokes that 'it is impossible to conceive that the earth can move through space without communicating some motion to the aether which surrounds and pervades it'.[54] Furthermore, he showed in two papers[55] that if $udx + vdy + wdz$ were exact for the ether's motion, then light would travel in a straight line through the moving ether no matter what the speed of the ether with respect to the surface of the earth. When a wave such as PQ in figure 6.3 is tilted by the ether's motion to the position of P_1Q_1, it is propagated in a new direction within the ether. It is propagated in the direction of its new normal rather than in that of the z-axis. But, at the same time, the motion of the ether transports P_1Q_1 to the right in figure 6.3. Moreover, for any velocity of the ether at the earth's surface, Challis showed that these two influences on the path of P_1Q_1 balance each other exactly. Consequently, no matter with what percentage of its own speed the earth were to drag the ether along with it, the tilted wave would travel a rectilinear course along the z-axis. And, so long as light waves traversed a straight line, there was no contradiction between the undulatory theory and the standard explanation of aberration or, more exactly, Challis's version of it: 'There is nothing improbable in the supposition [that $udx + vdy + wdz$ is an exact differential]: it saves the undulatory theory; but I must protest against its being considered necessary for the explanation of the aberration of light.'[56]

Admitting the validity of this portion of Challis's research, Stokes conceded that light would traverse a straight line and that, therefore, aberration could be explained whether the ether were at rest with respect to the earth's surface or not. Thus, in a paper published in July 1846, he pointed out that the orthogonal trajectory in figure 6.2 represented the changing orientation of the wave front and 'must not be confounded with the path of a ray of light'.[57] Furthermore, Stokes wrote, Challis's results showed that 'the phaenomenon of aberration allows us to suppose that the aether passes through the atmosphere and through the earth with any velocity, either constant, or varying from point to point, provided only $udx + vdy + wdz$ be an exact differential'.[58] In his letter to Powell in December 1846, Stokes further clarified this point:

> Prof. Challis *has* shown that the expression for the change in the direction of the normal obtained when udx + . . . is supposed to be an exact differential allows us to suppose the motion of the ether at the surface of the earth relatively to the earth to be of any amount. . . .
>
> The result of an investigation on the undulatory theory is that a ray of light *will not* be propagated along a right line if the motion of the ether be such that udx + vdy + wdz be not an exact differential. When udx + . . . is an exact differential it so happens that the sideward motion of a ray due to the tilting of the plane of the wave, and consequent attraction, in the *direction* of propagation is exactly compensated at each point by the motion of translation of the ether in w^h the ray is propagated.[59]

Stokes agreed with Challis, therefore, that it was not necessary to suppose the ether at rest with respect to the earth's surface to explain aberration. Moreover, Challis had evidently caused Stokes to alter his view of aberration. It was no longer the changed orientation of the wavefront, but the rectilinearity of the path of a ray of light which was crucial. Whatever the change in orientation, it turned out, the ray's path would be straight. And that is what counted. Later recalling that he had not foreseen the rectilinearity of propagation when he wrote his first paper on aberration, Stokes noted that that was the 'tacit assumption' usually made in explaining aberration, 'and provided that be accounted for the rest follows as usual'.[60]

Nevertheless, Stokes continued to assume that the ether was, in fact, at rest with respect to the earth's surface. Writing to Powell he remarked:

> I believe Prof. Challis and I are agreed on all essential points relating to aberration. We differ I believe (1) as to our opinion whether it is the wire or the star that is seen out of its true place (2) as to considering it probable that the ether within the earth and close to the surface is at rest relatively to the earth, or that it is only carried with part of the earth's velocity.[61]

Stokes's reasons for supposing the ether to be at rest relative to the earth are not difficult to identify. First, he surmised that the earth's atmosphere might be able to ensnare ether intermingled with it.[62] Extremely tentative, this 'conjecture' at least supplied one feasible factor in determining the ether's status. More significant, he appears to have been quite impressed with South's 'remarkable experiment', and to have readily accepted the experiment's implications, including those pertaining to the interaction of the earth and the ether. Finally, results of experiments such as Arago's performed at the surface of the earth did, after all, 'follow immediately' from his own theory of aberration.[63] Therefore, permitted by the phenomenon of aberration to assume any value for the ether's velocity with respect to the earth's surface, Stokes was convinced by other considerations that the velocity was zero.

The jelly ether

Perhaps Stokes's most profound disagreement with Challis pertained to the nature of the ether. Assuming, as did Challis, that $udx + vdy + wdz$ had to be an exact differential, Stokes, in his July 1846 paper, broached the problem of determining what sort of a medium could actually possess this particular property. Challis had treated and continued to treat the ether as an ordinary fluid.[64] Imagining the case of a solid moving through an ordinary fluid, Stokes concluded that the resultant motion in the fluid would not render $udx + vdy + wdz$ exact.

> This appears to be due to two causes; first, the motion considered would probably be unstable in the part of the fluid behind the solid; and secondly, a tangential force is called into play by the sliding of one portion of fluid along another; and this force is altogether neglected in the common equations of hydrodynamics, from which equations the motion considered is deduced.[65]

To alleviate this problem one had only to accept the view of the ether that Stokes had already proposed in an earlier paper.[66] Then, the ether would behave as an elastic solid 'only for extremely small vibratory motions'. Furthermore, Stokes declared: 'If the motion be progressive, or not very small [vibrations], the aether will behave like an ordinary fluid. According to these views, therefore, the earth will set the aether in motion in the same way as a solid would set an ordinary incompressible fluid in motion.'[67]

In addition, because of the ether's solid properties, the irregularities arising in its motion 'will be propagated into space with the velocity of light', leaving a stable situation behind.[68] Because the irregularities were small, they were susceptible to propagation by the ether. Because they were propagated as soon as they occurred, they had no time to accumulate into large irregularities. Though 'identical in their physical nature with light', they might not produce the sensation of light either because they were too feeble, fell outside the visible spectrum or were too discontinuous.[69] Consequently, the fact that $udx + vdy + wdz$ was an exact differential necessitated the conclusion that the ether was not simply a fluid, but possessed solid properties as well.

In 1848 Stokes further elucidated his theory of the ether's nature by proposing his 'glue-water' theory. He imagined a large quantity of glue to be dissolved in water to form a stiff, elastic-solid jelly. Using the same appeal to the law of continuity that he had in 1845, he reasoned that a mixture containing very little glue would still have elasticity. It would behave as a fluid for large, slow-moving objects but would also support small, rapid transverse vibrations.

> Conceive now a medium having similar properties [as this diluted glue water], but incomparably rarer than air, and we have a medium such as we may conceive the aether to be, a fluid as regards the motion of the earth and planets through it, an elastic solid as regards the small vibrations which constitute light.[70]

Among other things, therefore, Stokes's theory of stellar aberration was a defence of the concept of an elastic-solid ether. Initially, by increasing the probability of an ether–earth interaction, the elastic-solid ether threatened havoc for the undulatory theory of light. Stokes argued that, on the contrary, these various ideas fit together quite harmoniously. His theory insisted on the necessity of attributing solid properties to the ether, while, at the same time, removing causes for apprehension about such an ether. He achieved this result by denying the distinction between fluids and solids. Specifically, he did it by equating two forces, the tangential force shown by South's experiment to exist in fluids and the resisting force of solids to transverse distortions. Pointedly noting the support that his theory provided the concept of an elastic-solid ether, Stokes wrote:

> It is certainly curious that the astronomical phaenomenon of the aberration of light should afford an argument in support of the theory of transversal vibrations [i.e. of an elastic-solid ether].
> Undoubtedly it does violence to the ideas that we should have been likely to form *a priori* of the nature of the aether, to assert that it must be regarded as an elastic solid in treating of the vibrations of light. When, however, we consider the wonderful simplicity of the explanations of the phaenomena of polarization when we adopt the theory of transversal vibrations, and the difficulty, which to me at least appears quite insurmountable, of explaining these phaenomena by any vibrations due to the condensation and rarefaction of an elastic fluid such as air, it seems reasonable to suspend our judgement, and be content to learn from phaenomena [including stellar aberration] the existence of forces which we should not beforehand have expected.[71]

Stokes and Challis apparently did agree on at least one basic point: that the ether was imponderable (i.e. not subject to gravitational attraction). In his lectures, Challis presented this as Newton's idea. 'Newton probably called it imponderable [because] he thought that the force of gravity itself might arise from the action of the ether so t[ha]t the ether w^d not be acted on by gravity being itself the origin of gravity.'[72] Years later, Stokes voiced the same argument. Distinguishing between ether and ponderable matter, he regarded ether as a material substance because it 'must possess that distinctive property of matter, inertia'. To any who would balk at the concept of space being filled with such matter, Stokes responded that the much greater difficulty was 'that of conceiving such an influence as that of gravitation to extend across an absolute void'. Newton himself abhorred that idea. If the undulatory theory required a luminiferous ether, Stokes asked, 'may it not be that that same substance forms, in some manner as yet unknown to us, the link of connexion whereby the sun is enabled to attract the earth, and keep it in its orbit?'[73]

With the tools of hydrodynamics and a realist concept of an imponderable ether, Stokes had moulded several initially disparate elements into a single theory. It united the phenomenon of aberration, the undulatory theory of light, the idea of an elastic-solid ether, the idea of a fluid ether and the assumption of the existence of an ether–earth interaction. Shaped very much within the Cambridge context, the theory conflicted with Challis's teachings but, in the end, was positively influenced by Challis's debating points. Though Stokes could see no way to put his and Fresnel's theories 'to the test of some decisive experiment', he confidently assumed that not many 'would be disposed to maintain Fresnel's theory, when it is shewn that it may be dispensed with, inasmuch as we would not be disposed to believe, without good evidence, that the ether moved quite freely through the solid mass of the earth'.[74] In later decades, consistent with the differences in their respective careers, it was Thomson moreso than Stokes who continued to develop the concept of the jelly ether. At the turn of the century, as Lord Kelvin, he was refashioning some old ideas.

Kelvin's cloud

In the spring of 1900, Kelvin delivered his well known lecture to the Royal Institution: 'Nineteenth Century Clouds over the Dynamical Theory of Heat and Light'. One of Kelvin's two clouds arose from the Michelson–Morley experiment of 1887, which presented crucial questions to those trying to understand the motion of matter through ether. Though possible answers lay in the suggestion by G F Fitzgerald and H A Lorentz that matter contracted as it moved with respect to the ether, Kelvin evidently regarded the physical cause of such a contraction to be unexplained and, therefore, the Michelson–Morley experiment to constitute a 'cloud'.[75] Delivered only a few years before Einstein's work on special relativity, Kelvin's lecture has rightly been seen as a kind of summary of one century's physics which pointed towards another's revolutionary concepts.[76] However, firmly rooted in Victorian ideas, the lecture also expressed the legacy of Stokes's jelly ether. For that concept, as pursued above all by Kelvin, had a long history. In fact, the 'Clouds' lecture was only one of a series of turn-of-the-century papers by Kelvin developing his latest, and radically different, insights into the jelly, or elastic-solid, ether. These papers thus provide a convenient focus for indicating the fate of Stokes's jelly ether and, also, for leading into the discussions of Kelvin's physical thought in Chapters 7 and 9.

Though Kelvin was always very careful in attributing any specific composition to the ether and though he changed his mind about what its probable composition was, there is no doubt that he accepted the ether's physical reality, at least from around 1850 to the end of his career. In 1854: 'Its existence is a fact that cannot be questioned.'[77] In 1884: 'We must not listen to any suggestion that we must look upon the luminiferous ether as an ideal way of putting the thing. A real matter between us and the remotest stars I believe there is, and that light consists of real motions of that matter. . . .'[78]

Moreover, for a long time Thomson applauded Stokes's explanation of planetary motions through the ether. He illustrated it with a long-running experiment, part of which is still in the Kelvin Museum attached to the Department of Natural Philosophy in Glasgow University. Shoemaker's wax was a solid that would transmit rapid transverse waves. But it was fluid enough to allow slow-moving objects to move through it. In the experiment he placed a slab of wax in water with corks beneath and bullets above. In about a year, the corks had risen through the wax while the bullets had sunk. In an 1884 lecture, Thomson explained that 'the motion of a cork or bullet, at the rate of one inch in 2,000 years, may be compared with that of the earth moving at the rate of . . . nineteen miles per second through the luminiferous ether'. His confident conclusion: 'When we can have actually before us a thing elastic like jelly and yielding like pitch, surely we have a large and solid ground for our faith in the speculative hypothesis of an elastic luminiferous ether, which constitutes the wave theory of light.'[79]

Stokes's theory received support from others at that time as well. Stokes himself, who had offered the theory in his professorial lectures of 1864 that Rayleigh attended, endorsed it as late as 1883:

> Nevertheless we are not absolutely driven to accept Dr Young's hypothesis; for there is as I have shown another way in which the law of aberration may be obtained; a way which though not free from difficulties exempts us from the necessity of supposing that the earth in its motion through the ether allows the ether to pass through it with absolute freedom.[80]

Maxwell wrote favourably of the idea in his article on 'Ether' for the *Encyclopaedia Britannica*.[81] When A A Michelson performed the first version of his experiment in 1881, which detected no motion of the ether with respect to the earth's surface, he thought he had confirmed Stokes's theory of aberration.[82]

But criticism lay ahead. In 1885 the Dublin physicist, G F Fitzgerald, strongly attacked the concept of the jelly ether. 'Although Professor Stokes seems to think that there is no contradiction in supposing the ether to be a jelly', Fitzgerald did.[83] He thought that such a thin jelly would be too rigid to allow matter to move readily through it but not nearly rigid enough to support the enormous strains required for electrical forces. In commenting on Thomson's Baltimore lectures of 1884, Fitzgerald had to protest 'strongly against Sir William Thomson's speaking of the ether as *like* a jelly'. Thomson was greatly mistaken in 'lending his overwhelming authority to a view of the ether which is not justified by our present knowledge'.[84] Like Thomson, Fitzgerald tried to reduce ether and matter to vortical motions in a fluid, but he disagreed that the resultant properties of ether would resemble those of jelly. In 1886 the Dutch physicist, H A Lorentz, rejected Stokes's analysis of a fluid ether flowing around a solid earth, arguing that no motion could meet Stokes's conditions.[85] When Michelson and Morley reported their results in 1887 they noted Lorentz's disproof of Stokes's theory. They also remarked that their experiment disproved Fresnel's theory and Lorentz's attempt to combine Fresnel's and Stokes's theories.[86]

In addition, these were the same years in which Thomson was losing confidence in his programme to explain ether and matter through vortex motions in a perfect fluid. He thus found in 1889 that he was compelled to accept *three* ultimate components of physical nature—ether, matter and electricity. He now found 'the difficulties are so great in the way of forming anything like a comprehensive theory, that we cannot even imagine a finger-post pointing to a way that can lead us towards the explanation.'[87]

He also regarded ether–matter interactions differently from only five years before. Sounding like Fitzgerald, he wrote of the problem of air molecules 'tearing through' an ether capable of supporting strong magnetic forces. 'How can it be that these prodigious forces are developed in ether, an elastic solid, and yet ponderable bodies be perfectly free to move through that solid?'[88] Thomson thought Fresnel and Stokes had done all that was humanly possible at the moment to explain aberration and the earth's motion through ether. 'It may be not beyond man's range to complete the solution—how the earth can tear through this elastic solid ether and yet the waves of light be propagated through it as they are. The aberration of light is still an absolute mystery.'[89]

By the turn of the century, such dilemmas had forced a revamping of Kelvin's ideas about ether and matter, thus bringing him closer to truth. In 1903, for example, he contrasted the 'artificial' models he had used in 1884 with 'the probably real details of ether, electricity, and ponderable matter, suggested in 1900–1903'.[90] Elasticity was now a basic property of a structureless ether, no longer caused by motions within a structured ether. Having to abandon the vortex theory that the ether's elasticity was a mode of motion, Kelvin was 'driven' to a different idea of ether: 'It seems to me indeed most probable that in reality ether is structureless; which means that every portion of ether however small has the same elastic properties as any portion however great.'[91] Ether and matter were now different entities, no longer unified at some deeper level. Ether was still compressible, but the compressibility was now tied to Kelvin's biggest shift in thought. Earlier physicists had not fully comprehended the ether as an elastic solid, Kelvin concluded, or they would have noticed 'the enormous difficulty presented by the laceration which the ether must experience' as matter moves through it.[92] Kelvin's solution to the problem of laceration was to assume that ether and matter can occupy the same space at the same time and, therefore, 'that ether is not displaced by ponderable bodies moving through space occupied by ether'.[93] Matter did exert a Boscovichian (i.e. action-at-a-distance) force on ether, so that ether within an atom of matter was condensed and rarefied into 'concentric spherical surfaces of equal density, but [with] the same total quantity of ether within its boundary as the quantity in an equal volume of free undisturbed ether'.[94] When accelerated, such an atom would produce longitudinal and transverse waves in the ether. When it exceeded the speed of light in ether, the effect would be similar to matter exceeding the speed of sound in air. Vastly smaller than these atoms, electrons were atoms of condensed (positive) or rarefied (negative) ether. Light waves were caused by electrons' vibrations.[95]

The ether's compressibility was thus essential to what Kelvin felt was a 'hopeful' approach to 'formidable difficulties'.[96] To Kelvin's mind it 'seems infinitely improbable that ether is infinitely incompressible'. This view 'appears more consistent with the analogies of the known properties of molar matter, which should be our guides'.[97] Therefore, because ether was compressible and almost surely extended throughout space without limit, it must also be imponderable. 'If ether extends through all space, then it is certain that ether cannot be subject to the law of mutual gravitation between its parts, because if it were subject to mutual attraction between its parts its equilibrium would be unstable, unless it were infinitely incompressible.'[98]

Consequently, in the closing years of his career, Kelvin continued to accept the concept of an elastic-solid ether, even in the face of sharp criticism from Maxwellians like Fitzgerald. To do so he had radically to change his mind about the ether's other properties. For example, Boscovichian action at a distance, previously rejected, now connected ether and matter. Contrary to his earlier thinking, which is discussed in Chapter 7, he now envisaged ether itself to be continuous, imponderable and different in kind from matter. Ironically, these were the properties that Stokes, in disagreement with Thomson, *had* attributed to ether. The compressibility of ether was one property, rejected by Stokes, which was part of both Kelvin's old and new viewpoints. Therefore, in an atmosphere of somewhat hostile Maxwellians and a mainly silent Stokes, Kelvin maintained the general Stokesian programme of research on an elastic-solid ether by adopting specific Stokesian properties, with the exception of incompressibility. The resultant ether, however, was far different from the pliable, gradually yielding jelly that Stokes had imagined in the 1840s. Nonetheless, in this its latest manifestation, Stokes's jelly ether lived until Kelvin's death in 1907.

Kelvin's new theories of the ether intertwined with his developing ideas about radioactivity, which are discussed in Chapter 9. What Kelvin called 'electrions', for example, consisted of rarefied ether, though that concept was not necessary for Kelvin's discussion of radioactivity. And, as we can see, the idea of a Boscovichian force producing concentric spheres within an atom was part of Kelvin's reasoning for some time before he invented the Boscovichian atoms of figures 9.6 and 9.7.

By concentrating on the 1840s and the period around 1900, this chapter has explored the chronological boundaries of the material discussed in the next chapter. Going back to the decades from the 1850s to the 1890s, Chapter 7 focuses on Kelvin's attempts to unify ether and matter through an understanding of their common composition. He sought comprehension of the jelly ether's internal structure along paths where Stokes was reluctant to follow.

1   J T Merz 1896–1914 *A History of European Thought in the Nineteenth Century* 4 vols (Edinburgh and London: William Blackwood and Sons) II 31–2n.

2   For a discussion of Fresnel and the resulting programme of research, see E Whittaker 1960 *A History of the Theories of Aether and Electricity* 2 vols (New York: Harper and Brothers) I 107–69 and J Z Buchwald 1981 'The Quantitative Ether in the First Half of the Nineteenth Century' in G N Cantor and M J S Hodge (eds) *Conceptions of Ether: Studies in the History of Ether Theories, 1740–1900* (Cambridge: Cambridge University Press) pp. 215–37.

3   T Young 1855 'Theoretical Investigations Intended to Illustrate the Phenomena of Polarisation: Being an Addition by Dr Young to M. Arago's "Treatise on the Polarisation of Light"' in G Peacock and J Leitch (eds) *Miscellaneous Works of the Late Thomas Young* 3 vols (London: J Murray) I 415. Young wrote this in 1823.

4   T Preston 1890 *The Theory of Light* (London: Macmillan and Co.) p. 29.

5   A Fresnel 'Sur l'influence du mouvement terreste dans quelques phénomènes d'optique' in H Senarmont *et al* (eds) *Oeuvres Complètes d'Augustin Fresnel* 3 vols (Paris: Imprimerie Impériale, 1866–70) II 627–36.

6   J Bradley 1728 'A Letter from the Reverend Mr. James Bradley, Savilian Professor of Astronomy at Oxford, and F. R. S. to Dr. Edmond Halley Astronom. Reg. &c. Giving an Account of a New Discovered Motion of the Fix'd Stars' *Phil. Trans. R. Soc.* **35** 646–8.

7   T Young 1804 'Experiments and Calculations Relative to Physical Optics', *Phil. Trans. R. Soc.* **94** 12–13.

8   F Arago 'Vitesse de la Lumière' in J A Barral (ed) *Oeuvres Complètes de François Arago* 17 vols (Paris: Gide and J Baudry, 1854–62) VII 562–8.

9   Fresnel 'Sur l'influence' p. 628. I have translated Fresnel's French with Charlotte Bruner's assistance.

10   *Ibid.* p. 632.

11   *Ibid.*

12   See, for example, F A Jenkins and H E White 1957 *Fundamentals of Optics* 3rd edn (New York: McGraw-Hill) 396–7.

13   Though hydrodynamics was part of Challis's course, Stokes's notes on Challis's lectures are confined to optics. (Stokes 'Prof. Challis's Lectures, Notes' *CUL Stokes Collection* PA35.)

14   *Memoir* I 8.

15   J Challis 1842 'Discussion of a New Equation in Hydrodynamics' *Phil. Mag.*, **20** 287–8.

16   Stokes 'On the Steady Motion of Incompressible Fluids' *MPP* I 1.

17   Challis to Stokes, 7 May 1842 *CUL Stokes Collection* C295.

18   Challis to Stokes, 20 August 1842 *CUL Stokes Collection* C297.

19   Stokes 1842 'On the Analytical Condition of the Rectilinear Motion of Fluids, with Reference to a Paper by Prof. Challis' *Phil. Mag.* **21** 297–300.

20   Stokes 'On the Theories of the Internal Friction of Fluids in Motion, and of the Equilibrium and Motion of Elastic Solids' *MPP* I 112–13.

21   J C Maxwell 1890 'Atom' in W D Niven (ed) *The Scientific Papers of James Clerk Maxwell* (Cambridge: Cambridge University Press) II 769.

22   Thomson to Stokes, 8 March 1852 *CUL Stokes Collection* K57. See the paragraph on Challis in I Grattan-Guinness 1985 'Mathematics and Mathematical Physics from Cambridge, 1815–40: A Survey of the Achievements and of the French Influences' in P M Harman (ed) *Wranglers and Physicists: Studies on Cambridge Mathematical Physics in the Nineteenth Century* (Manchester: Manchester University Press) p. 105.

23   Stokes 'Report on Recent Researches in Hydrodynamics' *MPP* I 157–87.

24   The notes were published in the *Cambridge and Dublin Mathematical Journal* from 1847 to 1849. (Stokes *MPP* II 1–7, 36–50, 221–42; Thomson *MPP* I 83–7, 107–12.)

25   Stokes 'On the Effect of the Internal Friction of Fluids on the Motion of Pendulums' *MPP* III 7. South had given Stokes permission to refer to

the experiment 'in any way you may think proper'. (J South to Stokes, 7 December 1850 *CUL Stokes Collection* S1199.)

26 Stokes 'On the Theories of the Internal Friction of Fluids' p. 76.

27 Stokes 'Report on Recent Researches' p. 187. As Stokes recorded in his report (p. 180), Challis's discussion of the pendulum had brought him into conflict with another Cambridge mathematician, G B Airy (*Phil. Mag.* for 1840 and 1841). Later, Stokes and Airy debated Challis on the theory of sound (*Phil. Mag.* for 1848 and 1849.)

28 B Powell 1846 'Remarks on Some Points of the Reasoning in the Recent Discussions on the Theory of the Aberration of Light' *Phil. Mag.* **29** 438–40.

29 Stokes to Powell, 10 December 1846 *CUL Stokes Collection* P533.

30 J F W Herschel 1845 'Light' in *Encyclopaedia Metropolitana* 26 vols (London: B Fellowes) IV 535. This article was written in 1827. In his widely read *Preliminary Discourse* of 1830, Herschel did not link the theory of transverse waves to any particular property of the ether. Neither did Whewell, even though he endorsed the idea of transverse waves even more strongly than Herschel. (*A Preliminary Discourse on the Study of Natural Philosophy* (New York and London: Johnson Reprint, 1966) §§289–90; W Whewell 1847 *History of the Inductive Sciences, from the Earliest to the Present Time* 2nd edn 3 vols (London: John W Parker) II 438–501; and W Whewell 1847 *The Philosophy of the Inductive Sciences, Founded upon Their History* 2nd edn 2 vols (London: John W Parker) I 315–17, 349–52; II facing p. 119 (Inductive Table of Optics).)

31 Stokes 'Challis's Lectures' (note 13).

32 G Green 'On the Laws of the Reflexion and Refraction of Light at the Common Surface of Two Non-Crystallized Media', 'Supplement to a Memoir on the Reflexion and Refraction of Light' and 'On the Propagation of Light in Crystallized Media' in N M Ferrers (ed) *Mathematical Papers of the Late George Green* (London: Macmillan, 1871) 243–69, 281–90, 291–311.

33 Stokes 'On the Theories of the Internal Friction of Fluids' p. 125.

34 *Ibid.* p. 127.

35 Stokes 'On the Dynamical Theory of Diffraction' *MPP* II 245.

36 Stokes to Powell, 10 December 1846 (note 29).

37 Stokes 'Dynamical Theory of Diffraction' p. 245.

38 *Ibid.* p. 257.

39 Lord Rayleigh, Notes on Stokes's lectures for 1864 *American Institute of Physics Niels Bohr Library* Microfilm of Rayleigh Papers box II reel 3.

40 Stokes 1845 'On the Aberration of Light' *Phil. Mag.* **27** 9; *MPP* I 134. The reprint omits the latter pages of the original but includes an 'Additional Note'. This paper was first read to the Cambridge Philosophical Society on 12 May 1845 (abstract in *Proc. Camb. Phil. Soc.* **1** (1843–63) 19) and secondly to the British Association meeting of 1845 (abstract in *Br. Assoc. Rep.* (1845) part 2 8). It was printed in the issue of *Phil. Mag.* for July 1845 (see first listing in this note, pp. 9–15) and, finally, was picked up by *Phil. Mag.* from the *Proc. Camb. Phil. Soc.* and printed in abstract form (*Phil. Mag.* **28** (1846) 62–3).

41   Stokes 1848 'On the Constitution of the Luminiferous Aether' *Phil. Mag.* **32** 343; *MPP* II 8.
42   Stokes 'Challis's Lectures' (note 13).
43   Stokes 'On the Aberration of Light' (note 40).
44   *Ibid.* and Challis 'On the Aberration of Light' *Br. Assoc. Rep.* (1845) part 2 8.
45   Powell 'Remarks on Some Points' (note 28) p. 425. After corresponding with Challis and Stokes, Powell added a supplement to his first paper. 'Note to a Former Paper on the Theory of the Aberration of Light' *Phil. Mag.* **30** (1847) 93–5.
46   Stokes 1846 'On the Constitution of the Luminiferous Aether, Viewed with Reference to the Phenomenon of the Aberration of Light' *Phil. Mag.* **29** 6–10; *MPP* I 153–6. The reprint omits the latter pages of the original.
47   Stokes 'On the Constitution of the Luminiferous Aether' (note 41).
48   Stokes 'On the Aberration of Light' (note 40) pp. 9–10.
49   *Ibid.* p. 10.
50   Challis to Stokes, 14 February 1846 *CUL Stokes Collection* C315.
51   Challis 1845 'On the Aberration of Light' *Phil. Mag.* **27** 323.
52   Challis 1869 *Notes on the Principles of Pure and Applied Calculation; and Applications of Mathematical Principles to Theories of the Physical Forces* (Cambridge: Deighton, Bell, and Co.) p. xvii. His 1852 paper is 'On the Cause of the Aberration of Light' *Phil. Mag.* **3** 53–4.
53   Challis 'On the Aberration of Light' (note 51) p. 323.
54   *Ibid.* p. 323.
55   *Ibid.* and Challis 1846 'On the Aberration of Light, in reply to Mr. Stokes' *Phil. Mag.* **28** 90–3.
56   *Ibid.* p. 93.
57   Stokes 'On the Constitution of the Luminiferous Aether' (note 46) p. 9.
58   *Ibid.* p. 9.
59   Stokes to Powell, 10 December 1846 (note 29). In the two instances in which ellipses follow '*udx* + ', they are Stokes's.
60   Stokes 'Additional Note' of 1880 added to 'On the Aberration of Light' (note 40) *MPP* I 138–40.
61   Stokes to Powell, 10 December 1846.
62   Stokes 'On the Constitution of the Luminiferous Aether' (note 46) p. 9.
63   Stokes 'On Fresnel's Theory of the Aberration of Light' *MPP* I 147.
64   Stokes noted that Challis regarded the ether as a fluid in 'On the Constitution of the Luminiferous Aether' (note 41) p. 344. The reprint omits the reference.
65   Stokes 'On the Constitution of the Luminiferous Aether' (note 46) p. 7.
66   Stokes 'On the Theories of the Internal Friction of Fluids' (note 20).
67   Stokes 'On the Constitution of the Luminiferous Aether' (note 46) p. 7.
68   *Ibid.* p. 8.
69   Stokes 'On the Constitution of the Luminiferous Aether' (note 41) p. 346.
70   *Ibid.* p. 347.

71 *Ibid.* pp. 346–7.
72 Stokes 'Challis's Lectures' (note 13).
73 Stokes *Burnett Lectures* pp. 15–16. The same argument is made in a manuscript that was probably a draft for this lecture: Stokes 'Physical Optics' *CUL Stokes Collection* PA188.
74 Stokes 'On Fresnel's Theory' (note 63) p. 147.
75 Kelvin 1901 'Nineteenth Century Clouds over the Dynamical Theory of Heat and Light' *Phil. Mag.* **2** 7 and Kelvin 1900 'On the Motion Produced in an Infinite Elastic Solid by the Motion through the Space Occupied by It of a Body Acting on It Only by Attraction or Repulsion' *Phil. Mag.* **50** 197–8. Despite their publication dates, the papers were first presented, respectively, in April and July 1900. They are reprinted as appendices B and A, respectively, in *Baltimore Lectures* (1904).
76 See, for example, P M Harman 1982 *Energy, Force, and Matter: The Conceptual Development of Nineteenth-Century Physics* (Cambridge: Cambridge University Press) pp. 149–55. Problems associated with the equipartition theorem constituted Kelvin's second cloud. Stokes, too, expressed reservations about the equipartition theorem. (Stokes to Kelvin, 8 June 1895 *CUL Kelvin Collection* S494.)
77 Thomson 'Note on the Possible Density of the Luminiferous Medium and on the Mechanical Value of a Cubic Mile of Sunlight' *MPP* II 28.
78 Thomson 1884 *Baltimore Lectures* p. 5.
79 Thomson 'The Wave Theory of Light' *PLA* I 355. For a description of the Kelvin Museum see G Green and J T Lloyd 1970 *Kelvin's Instruments and the Kelvin Museum* (Glasgow: University of Glasgow Press). Thomson's bowl of wax for this experiment is still in the museum (p. 51).
80 Stokes *Burnett Lectures* p. 24. Rayleigh, Notes on Stokes's lectures for 1864 (note 39).
81 J C Maxwell 'Ether' in *Scientific Papers* (note 21) II 769.
82 A A Michelson 1881 'The Relative Motion of the Earth and the Luminiferous Ether' *Am. J. Phys.* **22** 129.
83 G F Fitzgerald 'On a Model Illustrating Some Properties of the Ether' in J Larmor (ed) *The Scientific Writings of the Late George Francis Fitzgerald* (Dublin: Hodges, Figgis, and Co., 1902) p. 153.
84 G F Fitzgerald 'Sir W. Thomson and Maxwell's Electromagnetic Theory of Light' in *Scientific Writings* p. 173. For a discussion of Fitzgerald's criticisms, including long quotations from his papers, see H Stein '"Subtler Forms of Matter" in the Period Following Maxwell' in *Conceptions of Ether* (note 2) pp. 315–20. For a discussion of Fitzgerald's theory of ether and matter see D M Siegel 'Thomson, Maxwell, and the Universal Ether in Victorian Physics' in *Conceptions of Ether* pp. 258–9. During the late 1890s Fitzgerald continued to criticise Kelvin in correspondence, some of which is printed in H I Sharlin 1979 *Lord Kelvin: The Dynamic Victorian* (University Park and London: The Pennsylvania State University Press) pp. 227–30 and *Thompson* II 1064–72.

85   Whittaker *Aether and Electricity* (note 2) I 387.
86   A A Michelson and E W Morley 1887 'On the Relative Motion of the Earth and the Luminiferous Ether' *Am. J. Phys.* **34** 341.
87   Thomson 'Ether, Electricity, and Ponderable Matter' *MPP* III 510. See also 'Motion of a Viscous Liquid; Equilibrium or Motion of an Elastic Solid; Equilibrium or Motion of an Ideal Substance Called for Brevity *Ether*; Mechanical Representation of Magnetic Force' *MPP* III 465.
88   Thomson 'Ether, Electricity, and Ponderable Matter' p. 510.
89   *Ibid.* p. 503.
90   Kelvin 1903 'On Electro-Ethereal Theory of the Velocity of Light in Gases, Liquids and Solids' *Phil. Mag.* **6** 442. On Kelvin's ideas *circa* 1900, see *Thompson* II 1076–83 and Stein '"Subtler Forms of Matter"' p. 330. Siegel summarises the late-nineteenth-century ether theories of Fitzgerald, Oliver Lodge and Joseph Larmor in 'Thomson, Maxwell, and the Universal Ether' pp. 258–9, 262–3. None was the same as Kelvin's.
91   Kelvin 'On the Motions of Ether Produced by Collisions of Atoms or Molecules, Containing or Not Containing Electrions' *MPP* VI 236.
92   Kelvin 'Nineteenth Century Clouds' (note 75) pp. 2–3.
93   *Ibid.* p. 3. See also 'On the Motion Produced in an Infinite Elastic Solid' (note 75) where Kelvin discussed the idea in greater detail.
94   Kelvin 'Nineteenth Century Clouds' p. 3. He called the force of matter on ether a Boscovichian force in 'Electro-Ethereal Theory' (note 90) p. 437.
95   Kelvin 1900 'On the Duties of Ether for Electricity and Magnetism' *Phil. Mag.* **50** 305–7 and 'Electro-Ethereal Theory' pp. 437–8.
96   Kelvin 'On the Duties of Ether' p. 307.
97   Kelvin 1902 'On the Clustering of Gravitational Matter in Any Part of the Universe' *Phil. Mag.* **3** 2.
98   *Ibid.* pp. 1–2.

7

Thomson's Aerial Ether

Scholars have located several landmarks in Thomson's enduring efforts to comprehend ether and matter. In 1896 Kelvin himself spoke of his 1846 paper, 'On a Mechanical Representation of Electric, Magnetic, and Galvanic Forces', as the beginning of a half century of 'fits of ether dipsomania, kept away at intervals only by rigorous abstention from thought on the subject'.[1] In that paper he showed that solutions to Stokes's equations for an elastic solid provided analogues for electrical, magnetic and galvanic activity. In an influential 1856 paper, drawing on Rankine's idea that heat resulted from microscopical rotational motions, Thomson employed such rotational motions to explain the Faraday effect.[2] In an 1858 manuscript he wrote of 'the probable truth of the doctrine of the Universal Plenum' and speculated on the possibility of constructing a unified theory of physical phenomena in terms of solid rotating particles embedded within a universal fluid medium.[3] In 1867 he replaced the solids with his well known vortex atoms, and the vortex-atom theory lasted until the mid-1880s.[4] Using 'rude' mechanical models, Thomson's Baltimore lectures of 1884 dealt with relations between ether and matter and presented an optimistic assessment of how much was understood. By the 1890s, however, no doubt largely due to his abandonment of vortex atoms, he was more pessimistic. Around 1900, as discussed in Chapter 6, he began optimistic pursuit of still another theory of the ether.

Most of this story is familiar to Thomson scholars, having been discussed from the time of Thomson's scientist-biographers—G F Fitzgerald, Andrew Gray, Joseph Larmor and S P Thompson[5]—to recent years when historians of physics have been writing on Thomson. Largely overlooked, however, has been a brief, easily missed passage in a paper published by Thomson in 1854 on the density of ether. In conjunction with some of Thomson's correspondence with

Stokes from that period, this passage throws new light on Thomson's basic views of ether and matter. In revealing his conviction that ether and air were the same, the passage helps us see: (1) that 'Boscovich's theory' may well have played an important role in Thomson's thoughts about the ether; (2) that Thomson's 1858 'Universal Plenum' was not continuous; (3) that in 1858 he was not introducing elasticity as a primary property of the ether; (4) that the kinetic theory of gases caused him to change his concept of ether; (5) that in 1867 he evidently did expect vortex motion to help explain the luminiferous medium as well as atoms of matter; (6) that his concepts of the ether dovetailed with his belief in the probable existence of longitudinal ether waves; (7) that that belief provided one ground for his rejection of Maxwell's electromagnetic theory of light; and (8) that he and Stokes differed on conceptual and methodological issues.

This chapter tries to bring out these points by examining Thomson's views of ether and matter and Stokes's responses to them from the 1850s into the 1890s. Thomson's ideas changed from his 1854 view that ether and air were the same to a view that ether was highly similar to air. Thomson often discussed his ideas with Stokes, and Stokes was often sceptical. Shaped over the decades, their differences emerged again in their reactions to Röntgen's discovery of x-rays in 1895. Regarding such issues, Stokes was now a more cautious realist than Thomson.

The aerial ether

From as early as 1854 to as late as 1862 Thomson supported the concept of an aerial ether. He thought the earth's atmosphere extended throughout space, that this interstellar atmosphere was the medium that transmitted light, and that it was a 'fine-grained' medium, its constituent particles being vastly smaller and closer together than the length of a wave of light. He expressed his views in 1854 both in correspondence with Stokes and in a published paper, in 1857–8 in correspondence with Stokes, and in 1862 in a published paper and in correspondence with P G Tait. There are at least two men who may have influenced Thomson's thinking in this area—Faraday and Rankine.

In a speculative paper of 1846, Faraday suggested that both ether and matter were composed of Boscovichian point atoms, but that the forces or powers surrounding the ethereal point atoms were different— less numerous—than those surrounding material point atoms. He thought that a rarefied gas was the state of 'common matter' most closely approximating 'to the condition of the aether'. According to Gooding, Faraday was here opposing the concept that ether and matter were different in kind. Ether was just attenuated, ordinary matter which, through attenuation, had lost certain properties, like gravitation, which were possessed by matter in its familiar state.[6] By the 1852–3 session, Thomson was evidently telling his students that, according to Faraday, air was the medium which supported electrical activity. William Jack recorded the following in his notes on the lectures:

> Faraday says that it is the air [that is] put into a state of tension— bodies are pulled asunder by air[?]

*   *   *

> Faraday and Snow Harris say that elect[y] dont [sic] operate through conduction. Faraday considers that it was a tension of "air on" outside entirely—But electrical force acts everywhere—in every direction.[7]

Beginning in 1850, Rankine published a series of papers on his 'hypothesis of molecular vortices', which he applied especially to the phenomena of heat, but also light. 'Each atom of matter', Rankine proposed, 'consists of a nucleus or central point, enveloped by an elastic atmosphere.'[8] Rotation of the matter comprising the atmosphere constituted heat. The nuclei themselves, interconnected by their mutual attractions and repulsions, constituted the medium that transmitted the undulations of light and radiant heat. In 1853 he extended the theory to empty space. He imagined the nuclei stripped of their material atmospheres and 'distributed throughout all space'. Now each was conceived to oscillate round an axis, with the axes of all the nuclei being parallel to one another. Hence, for example, plane polarised light could be regarded as consisting of 'a small oscillatory movement of each atom round an axis transverse to the direction of propagation'.[9] Nuclei within ordinary matter were still conceived to be surrounded by atmospheres which 'loaded' the nuclei, thus affecting their transmission of light undulations. The nuclei were extremely small, perhaps merely points.

Both Faraday and Rankine raised the possibility that point particles were common to ether and matter, and both presented a less strict dichotomy between ether and matter than Stokes had, whether it was a total dichotomy or not. One might, for example, interpret Rankine's dichotomy between nuclei and atmospheres as a refined way of envisaging the ether–matter dichotomy, the nuclei constituting ether, the atmospheres ordinary matter. Or one might argue that Faraday's point particles with *different* powers actually constituted a substance different in kind from ordinary matter. Nevertheless, the ideas of both Faraday and Rankine were alternatives to Stokes's views and were obviously part of the background to Thomson's speculations about the ether. Thomson was deeply aware of Faraday's work, and Smith has discussed the mutual influence of Rankine and Thomson on one another in the early 1850s in regard to the subject of heat. Smith argues that Thomson persuaded Rankine to place less emphasis on hypothetical entities, but that he himself was influenced by Rankine's ideas in his own attempts to relate heat to microscopic rotatory motion.[10]

Whatever the exact influences on Thomson's thought, he evidently shifted from a view like Stokes's, that ether and matter were different in kind, to his new view sometime between 1850 and 1854, possibly as early as 1852. I interpret statements recorded by one of his students in the 1849–50 session to reflect a Stokes-like position. After emphasising the similarities of light and radiant heat, Thomson said: 'Is it possible for heat to pass where there is no matter. Surely for it comes from the sun thro' vacant space.'[11] Regarding the undulatory theory, he declared:

> Light may be matter in motion. We can't say that it is undulations of the air for light travels from the planets where there is no air. It is absolutely necessary to suppose a medium. We must suppose a kind of matter which is affected only by light.[12]

Considering both of these statements, it appears that Thomson thought that interplanetary space contained no ordinary matter but that it did contain another kind of matter, ether, which could transmit the undulations of light and radiant heat.[13] Certainly, he did not yet regard air as the medium of light. But he may have by 1852. Jack's lecture notes record not only Thomson's statements regarding Faraday on air and electricity, but also that Thomson may have adopted the idea of an interplanetary atmosphere by the autumn of 1852. After recording Thomson's discussion of the undulatory theory of light, Jack noted the following: 'An ether has [been] assumed—a luminiferous ether—No proof has been given of sudden cease of air. It [is] more probable [that there is] matter [than that there is not].'[14] In 1854 Thomson published the view himself.

In the late winter and spring of 1854 Thomson was utilising the dynamical theory of heat to investigate both the amount of energy emitted from the sun and the density of the luminiferous ether required to transmit that much energy. The amount of energy transmitted by the ether was a function of its density and of the speed with which its parts vibrated in transmitting the energy from the sun. Thomson wrote to Stokes on 2 March:

> The mechanical value of the sun's heat per sq. foot at the earth's distance is about 84 foot lbs per second. If you calculate from this according to any tenable supposition as to the velocities of the vibrating particles (e.g. circularly polarized light with $\frac{1}{100}$ of the veloc. of prop^n, as the velocity of the particles), you will be astounded with the greatness of the density of the luminiferous medium required to produce the mech^l eff^{ts}. If for instance there were only 1 ft lb per sec^d produced per square foot, and if the mean vel. of the particles be $\frac{1}{100}$ of the veloc. of light w^h is about 1000000000 ft per sec^d, the mass of vib^g particles in a cub. f^t would have to be $\frac{32 \cdot 2}{10^{23}}$ lbs. Now the density of the air in interpl. space, *if the temperature were uniform from the surf. of the earth upwards*, would be only some $\frac{1}{10^{230}}$ of the dens at the surface of the earth. *What is the lum^s medium then?* There must be matter in interplanetary space perhaps 10^{200} times as dense as the air w^d be on that hypoth.[15]

In reply, Stokes thought there was nothing 'unreasonable' in Thomson's value for the ether's density. 'But what principally struck me was the magnitude of the excursions.' Thomson's estimates meant that the amplitude of ethereal vibrations would have to be about one-sixth of a wavelength, making the amplitude very large, not very small as some thought, 'relatively to the particles of ponderable matter'. However, Stokes did not agree about the extent of the earth's atmosphere: 'I am altogether sceptical about the existence of air in the planetary spaces, but if it do exist I have no confidence in the truth of Boyle's law when pushed to such limits.'[16] Answering Thomson's subsequent question, 'How can you think the air stops?',[17] Stokes explained that, on Thomson's assumption, the effect of the sun's great mass would mean that Venus and Mercury 'must be circulating in an atmosphere of tremendously great density; and therefore offering an enormous resistance' and, second, that the atmospheric pressure at the earth's surface would vary significantly as its distance from the sun varied in its elliptical orbit around the sun. 'In the face of such perfectly extravagant results, will any reasonable supposition respecting the temperature or motion of this supposed common atmosphere set matters to rights?'[18]

Thomson thought so. He wrote to Stokes on 20 April:

> I have no doubt that the difficulty you show regarding the earth's atmosphere when at diff^t distances from the sun will be explained by taking into acc^t the centrifugal force due to the revolution round the sun of a portion of it carried round with the earth, or else, by considering w^h is very probable, the whole interplanetary atmosphere to be revolving round the sun. I am much disposed to go back to the Vortices,

differing only from Descartes' in being dragged round by the planets
instead of drag[g]ing them round.[19]

In his paper read to the Royal Society of Edinburgh on 1 May,
Thomson mentioned the problem of interstellar air's having sufficient
density but, nevertheless, in an inconspicuous parenthetical phrase,
endorsed the equivalence of the luminiferous ether and the interstellar
atmosphere.

> That there must be a medium forming a continuous material com-
> munication throughout space to the remotest visible body is a funda-
> mental assumption in the undulatory Theory of Light. Whether or not
> this medium is (as appears to me most probable) a continuation of our
> own atmosphere, its existence is a fact that cannot be questioned. . . .[20]

As James has pointed out, Thomson's concept of an interstellar
atmosphere fits with his theory of solar heat in 1854. That theory
required a solar atmosphere which James thinks Thomson identified
with dense ether concentrated around the sun by the sun's gravitational pull.[21] Thus, the two large gravitational bodies, earth and
sun, would have acted similarly on the ponderable aerial ether.

On the question of the size of particles of ponderable matter
compared to the dimensions of waves of light, Thomson agreed with
Stokes.

> I chose 1/100 of the vel. of propagation for the vel. of the particles, so
> that the excursions might be somewhat small fractions of the wave-
> lengths. I almost think that, besides being (as I never doubted they are)
> enormous compared with the intervals betw. the atoms of matter, they
> must be sometimes much greater than the wave length.[22]

That is, the amplitude could be a 'small fraction' of the wavelength
but still enormous compared to atomic dimensions.

The common view of Stokes and Thomson was evidently especially
important for Thomson's theory of the aerial ether. Consider, for
example, Stokes's statement that the ether had to be either a continuous medium or, if constituted of particles, effectively continuous so
far as light waves were concerned.

> The transmission of regular undulations, of which the period is
> arbitrary, at least within wide limits, requires us to suppose that the
> transmitting medium is either continuous or may be treated as such;
> that if it consist of ultimate molecules, or be otherwise heterogeneous,
> the number of intervals from molecule to molecule, or of deviations of
> one sign or another from an average homogeneity, shall be very great
> and as good as infinite within the length of a single undulation. . . .[23]

On this view, the evident continuity of lengths of light waves required a similar continuity in the medium transmitting them. Thus, if air were the medium of light, virtually an infinite number of its particles would have to be packed into the length of one light wave. That, it seems to me, is exactly what Thomson thought. Consider that, as explained in the next section, Thomson's abandonment of the concept of the aerial ether was tied to his realisation that the kinetic theory of gases required air to be 'coarse grained', to use his later terminology. It appears, therefore, that Thomson's view of air as being exceedingly 'fine grained' was essential to his concept of the aerial ether.

Before the kinetic theory exerted its influence, Thomson sought further to develop his unifying notion of ether and matter in a series of letters to Stokes between May and December 1857 and in a notebook entry of January 1858.[24] In the model presented there, he was especially trying to explain the Faraday effect which he had discussed in his 1856 paper.[25] In the Faraday effect the plane of transverse vibrations of light is rotated when the light is transmitted through heavy glass within a magnetic field, and Thomson had argued in his paper that the magnetic field must give rise to microscopic rotations which exist independently of the presence of light and which themselves cause the plane of vibrations of light to rotate. In the 1856 article he left it an open question whether ether and matter were continuous or particulate. In 1857–8, speculating on exactly what the rotations involved in the Faraday effect might be, he described a provisional model of particulate ether and matter in accord with his idea of an aerial ether. Though 'it does not seem probable that a complete theory of physical science can be founded on such a hypothesis',[26] Thomson clearly intended the model as a step towards a theory of the varieties of physical phenomena. He envisaged a perfect, frictionless fluid containing solid 'motes'. With this model he hoped to correlate observed physical phenomena with motions of the fluid and solids.

Rotation of the motes was crucial. Thomson thought that the translational motion of a large solid or of one of the motes themselves through the liquid 'will tend to generate, or will lose itself in, rotating motions of the motes in general'. This concept provided 'an illustration, so far as it goes perfect, of the generation of heat in a liquid by stirring it, and of the generation of heat in a liquid or solid by a current of electricity through it'.[27] He wrote to Stokes concerning the Faraday effect: 'If there was a preponderance of rotation [of the motes] in one direction about parallel axes, waves of transverse vibrations ⟨about⟩ in planes perpendicular to these axes would have Faraday's optical property of heavy glass . . .'.[28] The 'true nature of heat' consisted of motions, and it appeared to Thomson 'that these motions only want *setting* in that axial fashion, to give rise to Faraday's property'.[29]

Similar modes of thought underlay his statement in 1860: 'Whatever electricity is, it seems quite certain that electricity in motion IS *heat*; and that a certain alignment of axes of revolution in this motion IS *magnetism*.'[30] Clearly, Thomson was seeking not only to understand the Faraday effect, but to imagine a comprehensive material system underlying the phenomena of heat, electricity, magnetism and light.

In fact the properties of the luminiferous ether also depended on the disposition of the motes towards rotation. With a large number of motes, Thomson wrote to Stokes, 'the tendency of all motion among them & the liquid will be to rotation'. And a rapidly rotating mote in a perfect liquid 'will experience a highly intense repulsive action when brought near either fixed boundaries or other solids rotating or not'.[31] Not only did Thomson think that these microscopic processes would correspond to the observed elasticities of ordinary solids, fluids and gases, he also thought, as he wrote to Stokes, that 'an infinite number of such motes all rotating with great angular velocities will repel one another & keep up the kind of stability & relative stiffness, *required for luminiferous vibrations*'.[32] We can therefore see that Thomson did not equate the fluid with ether and the solid motes with matter. Rather, ether and matter each consisted of a combination of the fluid and motes. Consequently, Thomson was not introducing elasticity as a primary property of a continuous, fluid ether, but instead speculated that the elasticity of the luminiferous medium, like that of ordinary matter, depended on the rotational motion of the motes. Here, as in the title of his later paper, elasticity was viewed as a 'mode of motion'.[33] As we shall see, even when he gave up the idea of the oneness of air and ether, he retained the notion that they had similar structures.

Thomson continued to accept the idea of an interstellar atmosphere, and presumably the related idea of the aerial ether, at least into the early 1860s. In 1862, in a notebook which he and Tait posted back and forth to obtain one another's comments as they worked out plans for their *Treatise on Natural Philosophy* (1867), Thomson sketched some of his thoughts on Newton's laws of motion, writing on one page of the 'resistance of interstellar air' to the motions of the sun and the earth. In red ink, Tait encircled the word *air* and wrote on the facing page: '*That* is one of the great stumbling-blocks between us as joint authors. I can't rightly appretiate your idea of an unlimited atmosphere. I have seen *hints* of it in your papers, but *no reasoning*. Why not say matter?' Adjacent to Tait's remark, Thomson scrawled, 'Matter or medium if you please. But air seems to me simpler.'[34] The same year he wrote in the *Proceedings of the Literary and Philosophical Society of Manchester* that it appeared to him 'most improbable that there is any limit to our atmosphere'.[35] Published immediately after that paper was an extract of a letter from Thomson to Joule regarding experiments on contact electricity, about which Thomson, without giving any numerical values, declared: 'This of course will give a definite limit for the sizes of atoms, or rather, as I do not believe in atoms, for the dimensions of molecular structures.'[36] Whatever Thomson meant about not believing in atoms, it appears to have been this sort of research into atomic sizes—especially as disclosed by the kinetic theory of gases—that led him to abandon the aerial ether.[37]

The air-like ether

In at least three places, Thomson commented on his rejection of his statements of 1854. In 1884 he wrote: 'I then thought that the medium must be a continuation of our atmosphere. I could not say any thing like that now.'[38] In 1899: 'Not so now. I did not in 1854 know the kinetic theory of gases.'[39] In a 1903 annotation on his letter to Stokes of 9 March 1854: 'I now believe they [i.e. ethereal vibrations] are enormously *small* compared with the centres of neighbouring atoms in solids, liquids, and gases. In 1854 I believed sizes of atoms, and the smallest distance between centres of atoms in solids and liquids, to be vastly smaller than we now know them to be.'[40] As these statements indicate, Thomson evidently abandoned his concept of the aerial ether because the kinetic theory of gases showed that particles of matter were much larger than he had previously supposed them to be.

The kinetic theory had been developed far enough by Clausius and Maxwell during the 1860s that George Johnstone Stoney, for example, could publish papers in the late 1860s relevant to our topic. Utilising the kinetic theory of gases, Stoney calculated that in a vacuum obtainable in the laboratory there would be only about sixty molecules of gas per wavelength of orange light.[41] He further argued that in the upper regions where 'the atmosphere is both very cold and very attenuated',[42] the earth's gravitational attraction on gas molecules would produce a net downward movement which would limit the extent of the atmosphere. Whether Thomson knew the work of Stoney and others or not, therefore, the kinetic theory of gases, which Thomson later called 'the most decisive and discriminating method of estimating the size of atoms',[43] was during the 1860s challenging his earlier ideas.

Thomson published a short note in *Nature* in 1870 on the size of atoms and returned to the subject the following year in his presidential address to the British Association for the Advancement of Science. Evidently reflecting the history of his own thinking, Thomson wrote in *Nature* that because of difficulties in conceptualising the atom and its properties, 'chemists and many other reasonable naturalists of modern times, losing all patience with it, have dismissed it to the realms of metaphysics, and made it smaller than "anything we can conceive"'.[44] He expanded on this point in his presidential address. The conclusions that he had reported in *Nature* the year before, he said, had been reached in ignorance of the similar findings of Loschmidt and Stoney.

> The triple coincidence of independent reasoning in this case is valuable as confirmation of a conclusion violently contravening ideas and opinions which had been almost universally held regarding the dimensions of the molecular structure of matter. Chemists and other naturalists had been in the habit of evading questions as to the hardness or indivisibility of atoms by virtually assuming them to be infinitely small and infinitely numerous. We must now no longer look upon the atom, with Boscovich, as a mystic point endowed with inertia and the attribute of attracting or repelling other such centres with forces depending upon the intervening distances (a supposition only tolerated with the tacit assumption that the inertia and attraction of each atom is infinitely small and the number of atoms infinitely great).[45]

As Thomson explained in *Nature*, the kinetic theory of gases (plus considerations of optical dispersion, contact electricity and capillarity) indicated that atoms were much larger than previously thought, around 10^{-8} centimetres in diameter. To help his readers 'form some conception of the degree of coarse-grainedness indicated by this conclusion', he pointed out how large the molecules in a drop of water would be if the drop were the size of the earth: 'The magnified structure would be coarser grained than a heap of small shot, but probably less coarse grained than a heap of cricket-balls.'[46] Since a wavelength of light is around 10^{-5} centimetres, Thomson's 'coarse grained' matter involved particles about a thousand times smaller than one wavelength.

The references in the long quotation in the previous paragraph to Boscovich and to certain 'universally held' ideas lead me to conclude that these conceptual developments of *circa* 1870 constituted an important episode in Thomson's continuing reflections on 'Boscovich's theory'. Nichol had taught Boscovich's theory, and point particles were employed in both Faraday's and Rankine's theories of ether and matter. Insofar as 'Boscovich's theory' meant forces acting at a distance, Thomson had by 1870 rejected it in favour of Faraday's approach. As Thomson recollected the growth of his own early thought, it was Boscovich's theory, as learned from Nichol, with which Faraday's rejection of action at a distance had to compete.[47] Insofar as 'Boscovich's theory' meant a force curve and infinitely small particles of matter, however, Thomson apparently accepted a modified version of it into the 1860s. For example, in a lecture to his natural philosophy class of 1863, Thomson discussed Boscovich's theory along the same lines that John Leslie had in 1804. Leslie strongly endorsed Boscovich's idea of a force curve, and Thomson explained: 'The force between the particles varies at different distances, repulsive within a certain limit and attractive beyond it. At remote distances it will merge into a great law of attraction.'[48] Also like Leslie, Thomson rejected Boscovich's idea of *infinitely* small atoms.

> Every body may be divided into parts, each of these parts again subdivided; nor has any limit been yet found to this repeated process of decomposition. Each separate portion likewise retains all the properties of the mass from which it was detached. ⟨Hence⟩ But though imagination represents an interminable series of subdivisions, such cannot be the actual constitution of nature, w^h is always defined by number and measure. We may therefore infer the existence of certain ultimate portions ⟨of⟩ or atoms, endued with the very fewest qualities, but w^h by their various combinations constitute the particles of matter, or form the corpuscular composition of bodies.[49]

Given that Thomson felt that modifying Boscovich's theory was an appropriate way to formulate a theory of matter in the early 1860s and given the 'universally held' ideas that he felt the need to reject in 1870, my own guess is that Thomson derived his early concept of matter from Boscovich's. For Thomson, there existed very small (less than 10^{-8} centimetres but not quite zero) particles of matter crowded together in virtually unlimited numbers.

Supporting this interpretation is Thomson's emphasis in mid-century lectures on the extreme smallness of particles of matter. In 1849 he stated: 'Certain odours such as musk show the very great divisibility of matter. One grain of musk must be divided into 320,000,000,000,000 of parts.' Reflecting on the size (1/300 000 of an inch) of microscopic organisms and the even greater smallness of their ultimate parts, Thomson declared: 'It is utterly impossible to say with truth that such is the atoms of which every thing is composed. For whatever size we fix as the size of the atoms we may be sure they are far more minute.'[50] In the autumn of 1852 Thomson repeated these examples and added others. The colours of light reflected from a thin film of a soap bubble showed that the film could reach a thickness 'considerably less than the wave length of light—thinner than the 1/400000 inch'. The colouring property of a certain dye 'is retained by particles not more than 1,30000000 inch'. Thomson summarised: 'This proves minute *division*, not *divisibility*. Probably matter is infinitely divisible.'[51]

Thus, it may be that Thomson's concept of matter in the early 1850s *was* Boscovich's insofar as the dimension of the particles was concerned. If so, he had modified the concept by the early 1860s, rejecting the *infinite* divisibility of matter even though no limit had 'been yet found to this repeated process of decomposition'. By 1870 certain scientific developments had provided a limit. If this reconstruction is substantially correct then Boscovich's theory, though modified and even rejected in various ways during the 1850s and 1860s, was nevertheless the starting point underlying Thomson's conceptualisation of the aerial ether, the material substratum of a unified physical nature.

In any case, it would appear that the conceptual shift from imagining air to be exceedingly fine grained to realising that it was actually quite coarse grained led Thomson to abandon the aerial ether sometime during the 1860s, or at least by 1870. His new view appeared most explicitly in a statement in 1884 in his Baltimore lectures:

> It seems probable that the molecular theory of matter may be so far advanced sometime or other that we can understand an excessively fine-grained structure and understand the luminiferous ether as differing from glass and water and metals in being very much more finely grained in its structure.[52]

Thus, the kinetic theory of gases did not cause Thomson to give up the idea of a structured ether. As before, he thought the ether possessed a fine-grained structure. It was just that now the ether could not be equated with coarse-grained air. The ether no longer was air; it was like air.

This interpretation aids our understanding of Thomson's well known theory of vortex atoms, first published in 1867 at about the time of his transition from an aerial ether to an air-like ether. As Knudsen has observed, the solid motes of the late 1850s were in 1867 'replaced by the famous vortex atoms'.[53] Drawing upon Helmholtz's theoretical hydrodynamics and Tait's demonstrations with smoke rings, Thomson transferred the rotational motion from the solid motes to portions of the fluid itself, thus picturing atoms as smoke-ring-like whirlpools of ether. Because both Thomson's aerial ether theory and his air-like ether theory portrayed the ether as possessing a structure, one would expect his theory of 1867 to have been intended to apply to the medium of light as well as to atoms of matter, even though Thomson did not then develop that side of the theory. Writing to Thomson in 1867 that he was 'beginning to comprehend a little about the ring vortices', Fleeming Jenkin grasped this point: 'I suppose you must have some kind of motion such as ring vortices even for the luminiferous ether, for I certainly cannot see how in an absolutely full fluid any kind of vibration or motion can be propagated. There is nothing to separate one part from the other.'[54] That Thomson would have responded affirmatively to this query is indicated by his introduction to a paper published in 1887.

> I have found something seemingly towards a solution (many times tried for within the last twenty years) of the problem to construct, by giving vortex motion to an incompressible inviscid fluid, a medium which shall transmit waves of laminar motion as the luminiferous aether transmits waves of light.[55]

Again, towards the end of his life, he recalled his much earlier conception that 'ether is a fluid presenting appearances of elasticity due to motion, as in collisions between Helmholtz vortex rings'.[56] Hence, in continuation of his earlier views involving rotating solid motes, in 1867 vortex motion provided Thomson not only with a theory of atoms but also with a suggestive insight into the source of the elasticity, or stiffness, of the luminiferous ether.[57]

Thomson's view of an air-like ether repeatedly surfaced in his Baltimore lectures of 1884. Convinced that air and ether were not identical, he usually spoke of a continuous ether, quite different from matter but interacting with it. Nevertheless, the already cited statement concerning the fine-grained structure of ether indicated a deeper speculation of ether and matter as essentially similar to one another, as differing in degree not in kind. The idea reappeared in comparisons of ether and air. Noting in his second lecture, for example, that 'a solid mass must act relatively to the luminiferous ether as an elastic body imbedded in it of enormous mass compared with the mass of the luminiferous ether that it displaces', he thought that, whereas such an interaction was 'infinitely difficult to understand' in the case of glass or water, 'the luminiferous ether in air is very easily understood. We just think of the molecules of oxygen and nitrogen as if they were groups of jelly relative to the luminiferous ether. . . .'[58] In introducing a possible mode of producing longitudinal waves, Thomson asked his audience to imagine the generating body 'at any place in air, or in luminiferous ether (I cannot distinguish now between the two ideas). . . .'[59] In contrast to the usual notion that ether differed from ordinary elastic solids by not transmitting longitudinal waves, Thomson declared: 'It seems to me that there are exceedingly strong probabilities that there will be waves of condensation and rarefaction of the luminiferous ether.'[60] He thought such waves would be shown to explain the propagation of electrostatic forces, as suggested by his already cited paper of 1846, 'On a Mechanical Representation of Electric, Magnetic, and Galvanic Forces', in which he had shown 'that the static displacement of an elastic solid follows exactly the laws of the electro-static force'.[61] Finally, he stated that 'we have not the slightest reason to believe the luminiferous ether to be imponderable; it is just as likely to be attracted to the sun as air is'.[62]

Thomson's concepts of an aerial ether and of an air-like ether may seem to have seriously conflicted with available empirical evidence. We have already noted, however, Thomson's Cartesian-like answer to the 'perfectly extravagant results' pointed to by Stokes. Additionally, though, the obvious contrast between longitudinal sound waves and transverse light waves would seem to entail a similarly sharp contrast between the respective media, air and ether, which support them. At this stage we should recall Thomson's continued emphasis on condensational ethereal waves. Consider two more passages from the Baltimore lectures:

> If I knew what the magnetic theory of light is, I might be able to think of it in relation to the fundamental principle of the wave theory of light. But it seems to me that it is rather a backward step from an absolutely definite mechanical motion that is put before us by Fresnel and his followers to take up the so-called Electro-magnetic theory of light in the way it has been taken up by several writers of late. In passing I may say that the one thing about it that seems intelligable [*sic*] to me, I scarcely think is admissable [*sic*]. What I mean is, that there should be an electric displacement perpendicular to the line of propagation and a magnetic disturbance perpendicular to both. *It seems to me that when we have an electro-magnetic theory of light, we shall see electric displacement as in the direction of propagation*—simple vibrations as described by Fresnel with lines of vibration perpendicular to the line of propagation—for the motion actually constituting light. I merely say that in passing as perhaps some apology is necessary for my insisting upon the plain matter of fact dynamics and the true elastic solid as giving what seems to me the only tenable foundation for the wave theory of light in the present state of our knowledge.[63]

> *   *   *   *   *

> I believe that the velocity of propagation of electrostatic force is the unknown condensational velocity that we are speaking of. I say "believe" here in a somewhat modified manner. I do not mean that I believe this as a matter of religious faith, but rather as a matter of strong scientific probability.[64]

Conceptualising the ether in accordance with his mechanical analogy of 1846, Thomson expected the ether to support longitudinal electrical waves, thus lessening the difference between air and ether suggested by any consideration restricted to sound and light. Moreover, the strength of Thomson's conviction—and thus its importance to his deepest physical speculations—is documented by his employment of longitudinal ethereal waves as a ground for not accepting

Maxwell's electromagnetic theory of light. Furthermore, as a by-product of this discussion, we can now recognise Thomson's opposition to Maxwell's theory as, at least in part, a natural consequence of Thomson's long-held conceptual framework of which the aerial and air-like ethers were key elements, thereby rendering a longitudinal ether wave a physical possibility.[65]

Cautious realists in disagreement

Even before abandoning vortices in the 1880s, Thomson never claimed anything like total success. He distinguished between illustrations and explanations and avoided confusing rough models with nature's reality. Nevertheless, for three decades his speculations carried overtones of significant progress being made in probing the truths of nature's underlying dynamical unity. Stokes did not see things quite this way. In the 1840s and early 1850s, he had already published a view of ether and matter different from Thomson's and in correspondence had discounted Thomson's idea of an interstellar atmosphere. Consistently more cautious and pessimistic than Thomson, Stokes continued to disagree with him in one way or another from the late 1850s into the 1890s.

In two letters of February 1858, for example, Stokes criticised Thomson's hypothesis of a perfect fluid containing solid motes. 'Without having a decided opinion either way', he wrote Thomson, 'I have always inclined to the belief that the motion of a perfect incompressible liquid, primitively at rest, about a solid w^h continually progressed, was unstable.'[66] The next day he described a particular case of fluid motion which led him to that conclusion.[67] Moreover, he questioned Thomson's basic view of the Faraday effect which had led Thomson into this area of investigation in the first place: 'I certainly am by no means clear that magnetic rotation must be due to motions going on independently of luminiferous vibrations.'[68] In their discussions of vortex motions in the early 1870s, Stokes—once again in cautious counterpoint to Thomson's optimism—declared, 'I confess I am sceptical about the stability of many of the motions which you appear to contemplate.'[69]

Their differences persisted into the 1880s. While Thomson was lecturing to his physicists in Baltimore, Stokes was in the midst of a three-year series of talks to a general audience in Aberdeen entitled *On Light*. The tone and substance of their respective remarks on the ether differed considerably. Thomson confidently proclaimed that 'we know more about it [ether] than we do about air or water, glass or iron', while Stokes warned that 'we must beware of applying to the mysterious ether the gross notions we get from the study of ponderable matter'.[70] Regarding aspects of the Faraday effect, Stokes stated, almost as though responding directly to Thomson's proposals, 'all these are questions concerning the true answers to which we can affirm nothing, though plausible conjectures may in many cases be framed'.[71] There were none of Thomson's longitudinal ethereal waves in Stokes's book, and he declared 'that the elasticity of the ether is of an altogether different nature from that of air'.[72] Whereas Thomson had declared it likely that ether would be gravitationally attracted by the sun, Stokes surmised that ether might be the cause of gravity and regarded ether as an imponderable substance not subject to gravitational attraction.[73]

They also presented contrasting discussions of possible interactions between matter and ether. Though Thomson no longer regarded air and ether as identical, and though he employed rude mechanical models in his Baltimore lectures, he nevertheless developed a vision of ether and air dynamically locked together into a vibrating system, the essence of which 'must be the true state of the case'.[74] It was not so much that Thomson actually claimed that air helped to transmit light waves, but that it was Stokes who cited empirical evidence in order to accentuate the fundamental differences between ether and matter and to underscore how little was understood of their interaction. Glass, for example, could not transmit light waves because its 'carefully determined' elasticity was only sufficient to support transverse waves with a velocity 'almost immeasurably smaller than the velocity required to account for the refraction of light on the supposition that it is the glass itself that vibrates'.[75] The conclusion was strengthened by experiments on light transmitted through air of various densities ranging from near that of a liquid to near that of an interstellar vacuum. 'When air is rare, and makes a near approach to what we call vacuum, but which we must now conceive as space filled with the luminiferous ether, it is clear that it must be by the vibration of the ether that light is propagated in it.'[76] Hence, if rarefied air were unsuitable for the propagation of light, then the continuous change observed as the density was increased to near-liquid air meant there was no discontinuous switch from the ether as the medium of light to ordinary matter (or to a combination of ether and matter) as the medium.[77] Though 'the presence of the ponderable molecules interspersed through the ether' probably affected the propagation of light,

> the precise mechanism by which this result is brought about we do not know. It is easy to frame plausible hypotheses which would account for the result, but it is quite another matter to establish a theory which will

admit of, and which will sustain, cross-questioning in such a variety of ways that we become convinced of its truth.[78]

By the 1890s Stokes was emphasising the mysterious quality of the ether even more than before. It was 'that mysterious medium',[79] 'a medium so mysterious in its nature'.[80] Not that Stokes necessarily held a totally different view of the ether at the end of his career than at the beginning. He still regarded it as a material substance which possessed inertia[81] and which 'does really exist'.[82] Rather, there was a different attitude towards the comprehensibility of the ether. In the 1840s, by using analogies linked with ordinary, ponderable matter, he had succeeded in solving difficult problems in a way that seemed to indicate that the ether could be regarded as a mechanical medium. A full half century later no deeper insight was available, the ether having withstood the onslaught of efforts like Kelvin's. Although the Stokes of the 1840s might reasonably have expected such efforts eventually to succeed, the Stokes of the 1890s emphasised, for example, that the ether's ability to exhibit both fluid and solid characteristics 'is a mystery which we do not understand'.[83] Stokes lumped the ether together with other mysterious entities as non-mechanical, or at least not-yet-mechanical, parts of nature. Though material, ether was not mechanical. Stokes did not dogmatically eliminate the possibility of some future mechanical explanation, declaring in the case of gravity: 'It is true that many have speculated on possible explanations of gravitation, which would reduce it to what I have called the mechanical properties of matter, but no one as yet has arrived at any satisfactory theory.'[84] Stokes's unmistakable emphasis, however, was not on the possibility of a future explanation, but on its improbability. As we saw in Chapter 4, the mysterious ether thus formed part of Stokes's theological attack on materialism and determinism.

In 1895 Röntgen's publication of his discovery of x-rays prompted an extended correspondence between Kelvin and Stokes, once again bringing out their differences. To be sure, Kelvin's statements in the 1890s tended to be more restrained than earlier, no doubt largely as a result of his having to give up his once-promising vortex theory. Thus in 1893, after so many decades of trying to comprehend ether and matter, he could comment on the importance of cathode-ray research 'if a *first* step toward understanding the relation between ether and ponderable matter is to be made . . . '.[85] Three years later he expressed the same sentiment in an oft-cited passage from his address during the celebration of his Jubilee as professor of natural philosophy at Glasgow: 'I know no more of electric and magnetic force, or of the relation between ether, electricity, and ponderable matter than I knew and tried to teach to my students of natural philosophy fifty years ago in my first session as Professor.'[86] Yet, as discussed in the previous chapter, Kelvin was still much more active than Stokes in seeking to solve the mysteries of ether and matter and retained theoretical–methodological differences with Stokes. Whereas Stokes by the 1880s seemed sceptical that studies of gross matter would reveal much about the ether, Kelvin declared in 1902 that it was 'the analogies of known properties of molar matter which should be our guides' in attributing properties—like compressibility—to the ether.[87]

In their correspondence on x-rays, Kelvin, as before, expected ether, like ordinary matter, to possess longitudinal waves. 'I feel strongly disposed to Röntgen's own supposition of condensational–rarefactional waves, but still I see tremendous difficulties.'[88] Stokes, on the other hand, wrote to Kelvin that 'normal vibrations in the ether did not fit in with my own speculations [on x-rays]'. Continuing, he explained:

> If the ether be compressible, the only examples of compressibility that we know anything about would lead us to attribute discontinuity to the ether, to suppose that it too is made up of discrete things, molecules shall I call them? But how do these 'things' act on each other? Is it by a sort of impact and rebound? That would lead us to attribute again structure and molecular constitution to these supposed molecules of ether. Is it by action at a distance? That leads us to the old alternative of action across a perfect void, or an ether of the second order to account for the supposed action of one on another of two molecules of the ether of the first order. Either supposition reminds me of the rhyme
>
> ". . . . . and dogs have fleas that bite 'em,
> And big fleas have little fleas, and so ad infinitum".
>
> My mind inclines me to scratch off the fleas, and rest in the idea of a plenum of incompressible ether, which our present knowledge does not authorise us in going beyond.[89]

Within a short time, experimental results had convinced both Kelvin and Stokes that x-rays could not be longitudinal waves. Less than two weeks after declaring to Stokes his disposition towards longitudinal waves, Kelvin wrote Stokes of his conversion to the view that x-rays were transverse waves.[90] Their initial responses to Röntgen's discovery, however, had reflected the contrasting theoretical views they had developed over the years, Kelvin once more seeking longitudinal ethereal waves, Stokes once more insisting on the differences between ether and matter in a statement that underlined the limits of 'present knowledge'.

Finally, Thomson's various speculations over the years pressed the related notions of continuity, unity and simplicity further than Stokes's did. As discussed in the previous chapter, Stokes's use of the principle of continuity united ordinary solids and fluids and was extended by analogy to an essentially optical ether which was sharply distinguished from ordinary matter. Probing with less restraint, Thomson sought the unity of light, heat, electricity and magnetism in the even deeper unity of ether and matter themselves. An underlying unity of superficially distinct areas of physical nature would have been supported by the many experimentally verified connections between different areas, by similar mathematical structure found in the various areas,[91] and, above all, by the principle of the conservation of energy. Thomson may also have had a theological basis for such unity, for he declared in 1860 that research was progressing towards 'a stage of knowledge . . . in which unity of plan through an inexhaustibly varied execution will be recognized as a universally manifested result of creative wisdom'.[92] Perfectly in harmony with such thinking, the concept of an interstellar atmosphere was, as Thomson had after all answered Tait, 'simpler'. Though Thomson lost some unity and continuity at one level when he discovered that air and ether were different from one another, his theory of vortices repaired the loss at a deeper level. His speculations in the late 1850s on nature at the ultimate level involved a dichotomy between fluids and solids. Despite the unity between ether and air, there was a discontinuity at the ultimate level between fluids and solids, because both ether and ordinary matter were composed of a fluid-solid medium. With vortex motion, however, even though ether and air were no longer identical, the substratum of both was a fluid possessing no solids, only various kinds of motion.

Consequently, though throughout their careers Stokes and Thomson shared a methodology of cautious realism, they certainly emphasised different aspects of it at different times. In the 1840s, dealing more directly with the luminiferous ether than Thomson, Stokes had seemed the less cautious realist. Later, neither doubted the real existence of the ether, and each from his own vantage point affirmed the need for care in deciding what was knowledge and what mere speculation with respect to the ether. But while Thomson optimistically tried almost to will nature into a coherent dynamical system of which longitudinal ethereal waves were a logical component, Stokes cautiously pointed out the imperfections of current physical theory, declaring—typically—that 'we must be content to learn [the ether's properties] by degrees, as they may be revealed by the study of the phenomena which are referable to actions of the ether'.[93] For Stokes, empirical differences between sound and light were all important in envisaging their respective media; for Thomson, such differences were to be transcended by speculative insight. Even in the 1890s the ether held more 'mystery' for Stokes than for Kelvin. In making a general point on scientific methodology, Stokes drew a contrast which in suitable moderation he must more than once have applied to Thomson and himself:

> We are not to expect to evolve the system of nature out of the depths of our inner consciousness, but to follow the painstaking inductive method of studying the phenomena presented to us, and be content gradually to learn new laws and properties of natural objects.[94]

1   Kelvin to G F Fitzgerald, 9 April 1896, in *Thompson* II 1065. Thompson 'On a Mechanical Representation of Electric, Magnetic, and Galvanic Forces' *MPP* I 76–80. The paper is dated 28 November 1846.

2   Thomson 1856 'Dynamical Illustrations of the Magnetic and the Helicoidal Rotatory Effects of Transparent Bodies on Polarized Light' *Proc. R. Soc.* **8** 150–8.

3   Ole Knudsen 1972 'From Lord Kelvin's Notebook: Ether Speculations' *Centaurus* **16** 47.

4   Robert Silliman 1963 'William Thomson: Smoke Rings and Nineteenth-Century Atomism' *Isis* **54** 461–74.

5   G F Fitzgerald 1899 'Biographical Sketch' in *Lord Kelvin: Professor of Natural Philosophy in the University of Glasgow, 1846–1899* (Glasgow: James MacLehose and Sons) pp. 1–29; A Gray 1908 *Lord Kelvin: An Account of His Scientific Life and Work* (London: J M Dent and Co.); J Larmor 1908 'William Thomson, Baron Kelvin of Largs, 1824–1907' *Proc. R. Soc.* **81** iii–lxxvi; and *Thompson.*

6   M Faraday 1855 'Thoughts on Ray-Vibrations' in *Experimental Researches in Electricity* 3 vols. (London: Taylor and Francis) III 448–9, 452; D Gooding 1980 'Faraday, Thomson, and the Concept of the Magnetic Field' *Br. J. Hist. Sci.* **13** 100–2; D Gooding 1981 'Final Steps to the Field Theory: Faraday's Study of Magnetic Phenomena, 1845–1850' *Hist. Stud. Phys. Sci.* **11** 252–3. I have profited from conversation with Dr Gooding on this topic.

7   W Jack, Notes on Thomson's lectures for 1852–53, *GUL* MS Gen. 130. Square brackets in quotations from Jack's notes indicate words that Jack wrote in shorthand. In this quotation I have added the question mark and *sic.*

8   W J M Rankine 1850 'On the Hypothesis of Molecular Vortices, and Its Application to the Mechanical Theory of Heat' *Proc. R. Soc. Edinburgh* **2** 276. See E M Parkinson 1975 'Rankine, William John Macquorn' *Dictionary of Scientific Biography* vol. XI (New York: Scribner) pp. 291–2.

9   W J M Rankine 1853 'General View of an Oscillatory Theory of Light' *Phil. Mag.* **6** 406.

10   C W Smith 1976 'William Thomson and the Creation of Thermodynamics: 1840–1855' *Arch. Hist. Exact. Sci.* **16** 253–61; C W Smith

1978 'A New Chart for British Natural Philosophy: The Development of Energy Physics in the Nineteenth Century' *Hist. Sci.* **16** 246–7; and C W Smith 1980 'Engineering the Universe: William Thomson and Fleeming Jenkin on the Nature of Matter' *Ann. Sci.* **37** 399–400.

11 W Smith, Notes on Thomson's lectures for 1849–1850, 29 January 1850 *GUL* MS Gen. 142.

12 *Ibid.* 2 November 1849.

13 Since Kelvin near the end of his life recalled his antipathy to imponderables during the 1840s, it is possible that in 1849 he regarded both ether and matter as ponderable substances, subject to gravitational attraction. (Kelvin 1902 'On the Clustering of Gravitational Matter in any Part of the Universe' *Phil. Mag.* **3** 2.) If so, the distinction between ether and matter would not be that between imponderable and ponderable. Smith interprets these lecture notes to mean that in 1849 Thomson already thought ether and matter differed only in degree. (Smith 'Engineering the Universe' (note 10) p. 397.) Of course, even if Thomson thought the difference was one of degree, this still constitutes an attempt to make a statement about the unobservable.

14 Jack, Notes on Thomson's lectures (note 7).

15 Thomson to Stokes, 2 March 1854 *CUL Stokes Collection* K64. Second set of italics added.

16 Stokes to Thomson, 7 March 1854 *CUL Kelvin Collection* S367.

17 Thomson to Stokes, 9 March 1854 *CUL Stokes Collection* K66.

18 Stokes to Thomson, 28 March 1854 *CUL Kelvin Collection* S369.

19 Thomson to Stokes, 21 March and 20 April 1854 *CUL Stokes Collection* K68.

20 Thomson 'Note on the Possible Density of the Luminiferous Medium and on the Mechanical Value of a Cubic Mile of Sunlight' *MPP* II 28.

21 F A J L James 1982 'Thermodynamics and Sources of Solar Heat, 1846–1862' *Br. J. Hist. Sci.* **15** 169. The solar atmosphere resisted the motion of meteors orbiting the sun, thus causing them to fall into the sun converting their gravitational potential energy into heat.

22 Thomson to Stokes, 9 March 1854 *CUL Stokes Collection* K66. In an annotation to this passage in 1903, Kelvin wrote that 'I cannot think how I *ever* imagined the excursions, in the most intense light or radiant heat, to be greater than *very* small in proportion to the wave length.'

23 Stokes *Burnett Lectures* pp. 19–20.

24 Knudsen 'Kelvin's Notebook' (note 3).

25 Thomson 'Dynamical Illustrations' (note 2).

26 Knudsen 'Kelvin's Notebook' (note 3) p. 47.

27 *Ibid.* p. 48.

28 Thomson to Stokes, 23 May 1857 *CUL Stokes Collection* K97. These Thomson–Stokes letters are also discussed in the detailed analysis of Thomson's interactions with his contemporaries in Smith 'Engineering the Universe' (note 10) especially pp. 400–2.

29 Thomson to Stokes, 17 June 1857 *CUL Stokes Collection* K98.

30 Thomson 'Atmospheric Electricity' *PEM* p. 224.

31   Thomson to Stokes, 17 June 1857.

32   *Ibid.* Italics added. He had earlier written to Stokes: 'I have been thinking too a great deal about the resultant pressures on solids rotating near one another in a perfect liquid. I can see clearly resultant repulsions in some cases and attractions in others, but I cannot yet make out an average repulsion of each from its neighbours when there are a great many scattered through the liquid, each rotating, and all in different directions and about non parallel axes. If I could, *I think I should have a very good medium for a mechanical illustration of light.*' (Thomson to Stokes, 23 May 1857. Italics added.) In the notebook entry he wrote: 'If the motes are in any way, by mutual repulsions whether in virtue of fluid pressure or of an occult quality, distributed and retained stably in or averaging near fixed positions with *the kind of rigidity required for undulations of light* (i.e. relative mobility with forces of restitution, but, comparatively reckoned, infinite resistance to crowding together, or rarefaction) and if on the whole their rotations are round axes parallel to one line', then he could demonstrate the Faraday effect. (Knudsen 'Kelvin's Notebook' (note 3) p. 49. Italics added.)

33   Thomson 'Elasticity Viewed as Possibly a Mode of Motion' *PLA* I 149–53. Lecture delivered in 1881. Compare Knudsen's statement in Knudsen 'Kelvin's Notebook' (note 3) pp. 43–4. Elasticity was still a property of the motes and would come into play if two motes collided, but elasticity was not an inherent property of the luminiferous medium. Kelvin regarded the motes' possession of elasticity as a serious weakness. (See Smith 'Engineering the Universe' (note 10) p. 402.) The luminiferous medium was continuous in the sense that it contained no empty spaces. But it was not continuous in the sense of being a completely homogeneous fluid. This latter sense of continuous is the one conveyed by existing literature on Thomson's view of the ether in the 1850s. Harman's superb summary of nineteenth-century physics, for example, states: 'In the late 1850s Thomson favoured the theory that the ether was to be envisaged as a fluid.' (P M Harman 1982 *Energy, Force, and Matter: The Conceptual Development of Nineteenth-Century Physics* (Cambridge: Cambridge University Press) p. 83. See also B G Doran 1975 'Origins and Consolidation of Field Theory in Nineteenth-Century Britain: From the Mechanical to the Electromagnetic View of Nature' *Hist. Stud. Phys. Sci.* **6** 135, 172, 178, 179, 185, 188 and D M Siegel 1980 'Thomson, Maxwell, and the Universal Ether in Victorian Physics' in G N Cantor and M J S Hodge (eds) *Conceptions of Ether: Studies in the History of Ether Theories, 1740–1900* (Cambridge: Cambridge University Press) pp. 245–6.)

34   The notebook still has two address labels on the cover, one addressed to Thomson and post-marked Edinburgh, 5 February 1862, the other addressed to Tait and post-marked Glasgow, 19 February 1862, and Edinburgh, 19 November 1862. (*CUL Kelvin Collection* NB47.)

35   Thomson 1862 'On the Convective Equilibrium of Temperature in the Atmosphere' *Proc. Lit. Phil. Soc. Manchester* **2** 172.

36 Thomson to Joule 1862 *Proc. Lit. Phil. Soc. Manchester* **2** 178. Thomson may not yet have had specific values in early 1862. Maxwell had written to him at the end of 1861: 'I shall be glad to know the max^m breadth of atoms. I suppose their length is not so easily limited, but if you could get a minimum breadth, you would go far to establish the existence of the atom.' (Maxwell to Thomson, 17 December 1861 in J Larmor (ed) 1937 *Origins of Clerk Maxwell's Electrical Ideas as Described in Familiar Letters to William Thomson* (Cambridge: Cambridge University Press) p. 39.)

37 Regarding belief in atoms, Thomson had written in his notebook entry in 1858 that the motes were not necessarily atoms because he thought the property of indivisibility or infinite strength was probably not possessed by any piece of matter. (Knudsen 'Kelvin's Notebook' (note 3) p. 47.) For a discussion of nineteenth-century research on atomic sizes see S G Brush 1976 *The Kind of Motion We Call Heat: A History of the Kinetic Theory of Gases in the 19th Century* 2 vols. (Amsterdam: North-Holland) I 75–8, 194–210.

38 Thomson 1884 *Baltimore Lectures* p. 200.

39 Thomson 1904 *Baltimore Lectures* p. 260n. This is a footnote dated 13 October 1899 to the parenthetical phrase of 1854 (see note 20).

40 Thomson to Stokes, 9 March 1854 (note 17).

41 G Johnstone Stoney 1868 'The Internal Motions of Gases Compared with the Motions of Waves of Light' *Phil. Mag.* **36** 141n, and 1869 'On the Physical Constitution of the Sun and Stars' *Proc. Roy. Soc.* **17** 14.

42 Stoney 'On the Physical Constitution' (note 41) p. 14.

43 Thomson 'The Size of Atoms' *PLA* I 220. Lecture delivered in 1883.

44 Thomson 1870 'The Size of Atoms' *Nature* **1** 551.

45 Thomson 'Presidential Address to the British Association, Edinburgh, 1871' *PLA* II 167.

46 Thomson 1870 'The Size of Atoms' *Nature* **1** 553.

47 *Thompson* I 20. In 1893, Kelvin stated that Faraday's theories had challenged and replaced the action-at-a-distance view successfully established by Boscovich. (Presidential address to the Royal Society of London, 30 November 1893 *PLA* II 538–41.)

48 D Murray, Notes on Thomson's lectures for 1862–1863, 5 March 1863 *GUL* MS Murray 326; J Leslie 1804 *An Experimental Inquiry into the Nature and Propagation of Heat* (London: J Mawman) p. 124. Murray's notes indicate that Thomson referred the class to Leslie for an explanation of Boscovich's 'curve of primordial action'. By this time, of course, Thomson did not regard attractions and repulsions as action-at-a-distance forces.

49 Murray, Notes on Thomson's lectures, 5 March 1863 (note 48); Leslie *Experimental Inquiry* (note 48) pp. 122–3.

50 W Smith, Notes on Thomson's lectures for 1849–1850, 7 November 1849 (note 11).

51 W Jack, Notes on Thomson's lectures for 1852–1853 (note 7).

52 Thomson 1884 *Baltimore Lectures* p. 10.

53   Knudsen 'Kelvin's Notebook' (note 3) p. 44.
54   Jenkin to Thomson, 23 May 1867 *GUL Kelvin Papers* J65.
55   Thomson 'On the Propagation of Laminar Motion through a Turbulently Moving Inviscid Liquid' *MPP* IV 308. He still was not completely successful, concluding 'that the most favourable verdict I can ask . . . is the Scottish verdict of *not proven*'. (*Ibid.* p. 320).
56   Kelvin 'On the Motions of Ether Produced by Collisions of Atoms or Molecules, Containing or Not Containing Electrions' *MPP* VI 236. He says there that he had abandoned the idea more than thirty years earlier (i.e. before 1877), thus raising a possible problem of chronology, for he was favourably inclined towards the idea of elasticity resulting from vortex motion in papers published after 1877. Perhaps he meant 20 years ago. In any case, his statement supports the view that he initially intended vortices to account for the elasticity of the ether as well as to explain properties of ordinary matter.
57   This somewhat modifies the standard view expressed in E Whittaker 1960 *A History of the Theories of Aether and Electricity* vol. I: *The Classical Theories* (New York: Harper and Brothers) pp. 294–5. A similar modification is suggested by statements in Siegel 'Thomson, Maxwell, and the Universal Ether' (note 33) p. 256, and Harman *Energy, Force, and Matter* (note 33) pp. 83–4.
58   Thomson 1884 *Baltimore Lectures* p. 28.
59   *Ibid.* p. 41.
60   *Ibid.* p. 43.
61   *Ibid.* p. 43.
62   *Ibid.* p. 207.
63   *Ibid.* pp. 5–6. Italics added. The 1904 edition removed the slight confusion in the sentence containing the italicised passage. In 1904 the first dash was replaced by *and*, the second dash by a comma, thus making clear that, unlike the longitudinal electric displacement, light waves were transverse. (Kelvin 1904 *Baltimore Lectures* p. 9.)
64   Thomson 1884 *Baltimore Lectures* p. 143.
65   Citing the passage from Thomson's *Baltimore Lectures* (1884) (pp. 5–6) which I have italicised, Andrew Gray briefly noted this objection by Thomson to Maxwell's theory. (Gray *Kelvin* (note 5) p. 255.) Larmor's and Thompson's biographies of Kelvin, however, have been much more widely used by modern scholars than Gray's. Gray's is not listed, for example, in J Buchwald 1976 'Thomson, Sir William' *Dictionary of Scientific Biography*, vol. XIII (New York: Scribner) p. 388. Thomson's rejection of Maxwell's theory is ignored in Larmor 'Kelvin' (note 5). Thompson discusses Thomson's rejection of Maxwell's theory and Thomson's view of longitudinal ether waves, but without explicitly connecting the two. For example, in quoting from the passage from Thomson's *Baltimore Lectures* (1884) (pp. 5–6), he stops before reaching the part I have italicised. (*Thompson* II 819; see also pp. 822–4, 831, 835, 836, 884, 885, 1021–6, 1041.) For more recent statements see Buchwald 'Thomson' (above) p. 385; Harman, *Energy, Force, and Matter*

(note 33) p. 100; and Kenneth F Schaffner 1972 *Nineteenth-Century Aether Theories* (Oxford: Pergamon) p. 69, where the passage from Thomson *Baltimore Lectures* (1884) (pp. 5–6) is quoted, omitting the part I have italicised. In a very recent essay, Knudsen quotes the italicised portion of the quotation from *Baltimore Lectures* (1884) and does discuss the importance of longitudinal ether waves in Thomson's response to Maxwell's theory. (O Knudsen 1985 'Mathematics and Physical Reality in William Thomson's Electromagnetic Theory' in P M Harman (ed) *Wranglers and Physicists: Studies on Cambridge Mathematical Physics in the Nineteenth Century* (Manchester: Manchester University Press) pp. 171–6.)

66 Stokes to Thomson, 12 February 1858 *CUL Kelvin Collection* S391.
67 Stokes to Thomson, 13 February 1858 *CUL Kelvin Collection* S392.
68 Stokes to Thomson, 12 February 1858.
69 Stokes to Thomson, 18 January 1873 *CUL Stokes Collection* NB21.57.
70 Thomson 1884 *Baltimore Lectures* p. 10; Stokes *Burnett Lectures* p. 80.
71 Stokes *Burnett Lectures* p. 169. For what Stokes did think was known, see *ibid.* pp. 85–8, 168–9.
72 *Ibid.* p. 76. Stokes did refer to longitudinal vibrations in the ether in his later Gifford lectures, saying that if they existed their speed would be infinite or nearly infinite. This was not a Kelvin-like attempt to explain physical phenomena, but an aid to his audience in conceiving 'of an intelligent will as pervading the whole universe'. (Stokes *Natural Theology* II 33.)
73 Stokes *Burnett Lectures* pp. 15–17.
74 Thomson 1884 *Baltimore Lectures* p. 28.
75 Stokes *Burnett Lectures* p. 79.
76 *Ibid.* p. 80.
77 *Ibid.* pp. 80, 97.
78 *Ibid.* p. 81.
79 Stokes 1896 'An Annual Address (Chiefly on the Subject of the Röntgen Rays)' *J. Trans. Victoria Inst.* **30** 14.
80 Stokes 1898 'On the Perception of Colour' *J. Trans. Victoria Inst.* **31** 254.
81 Stokes *Burnett Lectures* p. 15, and Stokes to A H Tabrum, 5 August 1900 *Memoir* I 84.
82 Stokes *Natural Theology* II 27.
83 *Ibid.* II 19.
84 *Ibid.* II 34–5. See also Stokes 1890 '"I:" A Lecture on the Immortality of the Soul' *The Family Churchman* (9 April) pp. 1–23, especially pp. 8–10.
85 Kelvin 1893 Presidential address *Proc. R. Soc.* **54** 389. Italics added.
86 Quoted in *Thompson* II 1072–3.
87 Kelvin 1902 'On the Clustering of Gravitational Matter in Any Part of the Universe' *Phil. Mag.* **3** 2.
88 Kelvin to Stokes, 12 February 1896 *CUL Stokes Collection* K309.
89 Stokes to Kelvin, 28 February 1896 *CUL Kelvin Collection* S501.
90 Kelvin to Stokes, 24 and 25 February 1896 *CUL Stokes Collection* K310.

91   See M N Wise 1979 'William Thomson's Mathematical Route to Energy Conservation: A Case Study of the Role of Mathematics in Concept Formation' *His. Stud. Phys. Sci.* **10** 49–83.
92   Thomson 'On Atmospheric Electricity' *PEM* p. 225.
93   Stokes *Burnett Lectures* p. 24.
94   *Ibid.* p. 25.

8

Stokes, Correspondent

Addressing section A of the British Association in 1862, Stokes declared: 'Any one who has worked in concert with another zealously engaged in the same research must have felt the benefit arising from the mutual interchange of ideas between two different minds.'[1] Not only a justification for the gathering of scientists in their sectional meetings, the statement also indicated the direction being taken by Stokes's own career. By then eight years into his long secretaryship of the Royal Society, he was routinely working 'in concert' with fellow scientists. In fact, the great majority of his own publications during the final half century of his career arose from collaborations with others.

In the process, Stokes amassed one of the great correspondences of the Victorian period—that grand era of correspondence, after the railway but before the telephone. By all accounts he was an assiduous writer of letters. One longtime correspondent, William Huggins, a fellow member of the Council of the Royal Society, put it this way:

> One of the most distinguishing characteristic qualities of Sir George, was the generous way in which he was always ready to lay aside at once, for the moment, his own scientific work, and give his whole attention and full sympathy to any point of scientific theory or experiment about which his correspondent had sought his counsel. Notwithstanding the many heavy duties resting upon him, his reply came nearly always without delay by an early post.[2]

Another, the optician Howard Grubb, recalled his reluctance to impose on Stokes, remembering the time Stokes wrote him five letters in a single day in response to an inquiry.[3] The Stokes Collection in the Cambridge University Library contains around 17 700 letters, the vast majority of which were written by others to Stokes. On the (perhaps conservative) assumption that Stokes wrote one letter for each received, Stokes would thus have written some 17 000 letters.

This chapter is a partial 'anatomy' of the Stokes correspondence. Picking up a theme from Chapters 2 and 3, it highlights Stokes's cooperation with non-mathematical physical scientists, especially William Crookes. Focusing on Crookes's research on the new physics of cathode rays and radioactivity, the chapter explores the nature and influence of the Stokes–Crookes collaboration. The collaboration reflected the primacy of mathematics, and of mathematical physicists, as the physical sciences became increasingly mathematized. The chapter tries to uncover Stokes's role in Crookes's cathode-ray research and to suggest why they were so successful in shaping a *British* view of cathode rays. Finally, it explores their correspondence on radioactivity, as part of the early stages of research in that field.

A Victorian correspondence network

When Maxwell published his first mathematical treatment of Faraday's qualitative idea of lines of force, he sent a copy to Faraday, initiating a brief correspondence between them. Faraday reported being 'at first almost frightened when I saw such mathematical force made to bear upon the subject, and then wondered to see that the subject stood it so well'.[4] Later, Faraday asked Maxwell more generally:

> When a mathematician engaged in investigating physical actions and results has arrived at his own conclusions, may they not be expressed in common language as fully, clearly, and definitely as in mathematical formulae? If so, would it not be a great boon to such as we to express them so—translating them out of their hieroglyphics that we also might work upon them by experiment.[5]

Faraday's well known responses to Maxwell epitomise pervasive factors within Victorian science—the distinctions and interactions between experimental and mathematical physics and between the experimenters and mathematicians themselves. The Thomson–Faraday and Thomson–Joule collaborations are two other prominent examples. Moreover, in such cases the mathematicians usually dominated. Maxwell's language was hieroglyphics to Faraday, not *vice versa*. Less famously, but more extensively, Stokes's voluminous correspondence largely reflects these same Victorian realities. Explaining a point in hydrodynamics to Faraday's eventual successor at the Royal Institution, John Tyndall, for example, Stokes could confidently write: 'I take for granted that you don't want to go into the mathematics.'[6]

Stokes's correspondence is not entirely devoted to matters of science, of course. There survives a substantial family correspondence—some 1060 letters from 73 relatives. This figure does not, however, include Stokes's correspondence with his father-in-law, Thomas Romney Robinson, director of the Armagh Observatory, which is more scientific than family. Most of the some sixty letters extant from Stokes to Robinson in 1877, for example, are detailed discussions of graphs of results in the latter's anemometer experiments, the method for graphing having been devised by Stokes and described in his appendix to one of Robinson's papers.[7] There is also a large religious correspondence. The Victoria Institute and its longtime secretary, Captain Francis W H Petrie, for example, account for 164 letters, while Stokes saved 267 from the Congregational minister, Edward White, his principal correspondent on conditional immortality.

Though large, these numbers still leave an enormous correspondence distributed among many scientific correspondents. Figure 8.1 provides an overview of the network of scientific correspondence centred on Stokes. The 40 men listed contrast with the 10 or so whom a similar graph for Kelvin would include. Since figure 8.1 selects by quantity of correspondence, it of course excludes some of Stokes's important correspondents: for example, James Challis (28 extant letters to Stokes), James Dewar (27), G D Liveing (36), James Clerk Maxwell (38), Edward Schunck (22),[8] and Arthur Schuster (16). Challis, Dewar, Liveing and Maxwell were Cambridge professors for varying periods, thus lessening the need for correspondence. Altogether, those in the figure account for over 4800 letters, and it will be noted that in the early 1870s Stokes was in correspondence with 34 of the 40.[9] In the five years from 1870 to 1874, they accounted for nearly 700 letters.

Much of Stokes's correspondence concerned the day-to-day business of scientific organisations: for example, most of that with the astronomer royal, G B Airy; that with the anatomist, Richard Owen, concerning the printing of his papers in the *Philosophical Transactions*; and that with fellow officers of the Royal Society. Stokes's leading role on the Solar Physics Committee from its inception in 1879 to his death obviously generated a large correspondence. Stokes greatly assisted J Norman Lockyer, for example, head of the Solar Physics Observatory in South Kensington. As Meadows has written: 'Stokes . . . advised on the adjustment and development of Lockyer's optical instruments, and during the seventies and eighties was often at South Kensington overseeing the work.'[10] Like Lockyer and Robinson, other experimental scientists also relied on Stokes's collaborative efforts.

Figure 8.1 (opposite and overleaf) Stokes's scientific correspondence. *Source:* D B Wilson 1976 *Catalogue of the Manuscript Collections of Sir George Gabriel Stokes and Sir William Thomson, Baron Kelvin of Largs in Cambridge University Library* (Cambridge: Cambridge University Library). Except for Robinson, Smithells and Kelvin, the figure records the number of letters *from* the correspondent to Stokes which are in the Stokes Collection, the assumption being that Stokes would have written about the same number in return. Stokes's letters *to* Robinson and Smithells have been completely enough preserved in the Stokes Collection that their number is given. Kelvin's 407 letters are all those from him to Stokes included in D B Wilson, *The Kelvin–Stokes Correspondence* (Cambridge: Cambridge University Press, forthcoming). The Kelvin-to-Stokes side of the correspondence has survived better than the reverse, for which there exist 249 letters. In addition to the number of letters, the figure records the span of time during which they were written, giving a rough idea of the 'density' of correspondence in each case. This can be misleading, for example in Tomlinson's case where most of the letters are concentrated in the years 1885 and 1886.

For example, despite his distinction in mathematics as an Oxford undergraduate and his Savilian professorship of geometry there, Baden Powell had devoted his chief researches to *experimental* studies of light and radiant heat in the 1820s. As we saw in Chapter 6, Powell corresponded briefly with Stokes in 1846 regarding his discussion of the Stokes–Challis conflict over stellar aberration. However, about three-quarters of the correspondence between Stokes and Powell stemmed from their collaboration of 1847–8. In the summer of 1847, Powell reported to Stokes observations he had made with an apparatus like that in figure 8.2. Here a hollow prism is filled with a liquid medium (M) into which is inserted a transparent plate (P). When light is transmitted through the arrangement, the resultant spectrum is interrupted by a series of dark bands. Stokes thought the phenomenon was related to phenomena already studied by W H Fox Talbot, David Brewster and G B Airy. In the ensuing correspondence, he swamped Powell with mathematical analyses of the phenomena as Powell tried to keep pace with experiments. Stokes predicted, for example, what Powell should observe with a certain modification of the original arrangement, and Powell duly confirmed the prediction. Their respective papers which followed from the joint investigation were quite different in scope, as Powell explained. Whereas Powell proposed 'to describe my own experiments, with the general application of the undulatory theory to the explanation of them', Stokes 'has gone extensively into the whole theory of these and other allied cases'.[11]

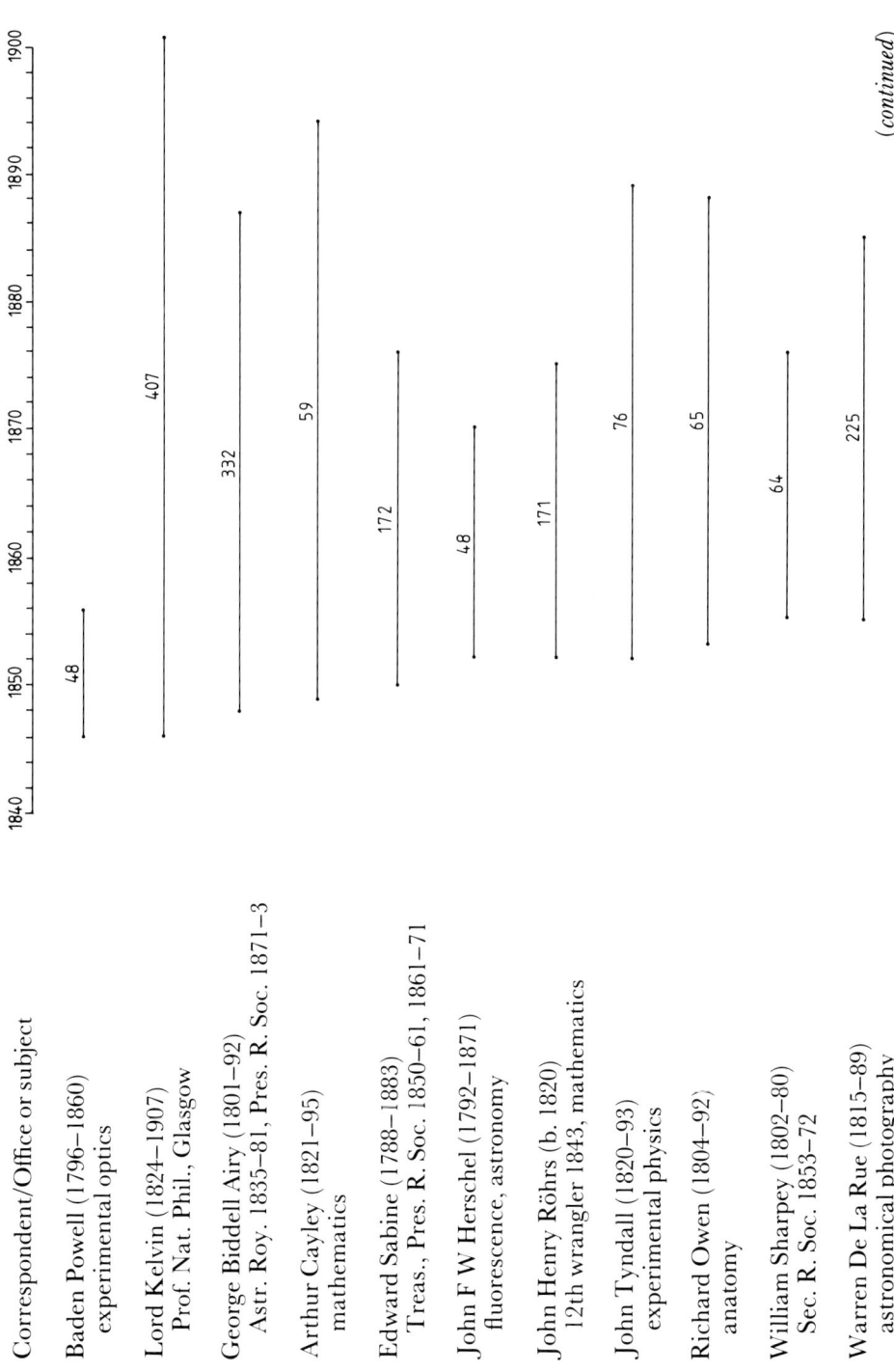

(*continued*)

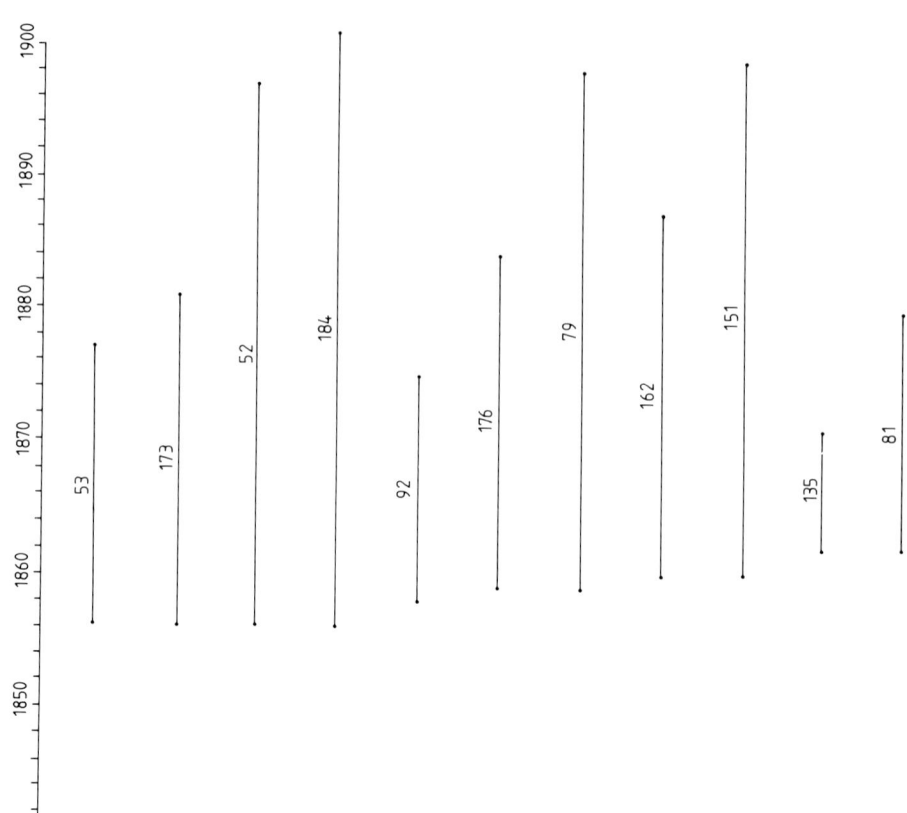

Correspondent/Office or subject

Taylor and Francis
     printers to R. Soc.

Thomas Romney Robinson (1792–1882)
     anemometers, father-in-law

Henry Enfield Roscoe (1833–1915)
     chemistry

William Crookes (1832–1919)
     experimental physics

Thomas Grubb (1800–78)
     optical engineer

Walter White
     Assist. Sec. R. Soc. 1861–85

George Gore (1826–1908)
     electrochemistry

Balfour Stewart (1828–87)
     Solar Physics Committee

Peter Guthrie Tait (1831–1901)
     Prof. Nat. Phil., Edinburgh

William Vernon Harcourt (1789–1871)
     chemistry of glass

Benjamin Collins Brodie (1817–80)
     chemistry

| 1840 | 1850 | 1860 | 1870 | 1880 | 1890 | 1900 |

Correspondent/Office or subject

William Spottiswoode (1825–83)
Treas., Pres. R. Soc. 1870–8, 1878–83   77

T H Huxley (1825–95)
Sec., Pres. R. Soc. 1872–81, 1883–5   77

William Huggins (1824–1910)
stellar spectroscopy   184

Herbert Tomlinson (1845–1931)
experimental physics   50

Fourth Earl of Rosse (1840–1908)
astronomy   53

Henry Clifton Sorby (1826–1908)
mineralogy, spectroscopy   82

Howard Grubb (1844–1931)
optical engineer   169

Robert Henry Scott (1833–1916)
Sec. Meteorological Council 1867–1900   166

Joseph Norman Lockyer (1836–1920)
solar physics   87

Henry Josiah Sharpe (1838–1917)
6th wrangler 1861, mathematics   165

William Alexander Ross
pyrology   218

(continued)

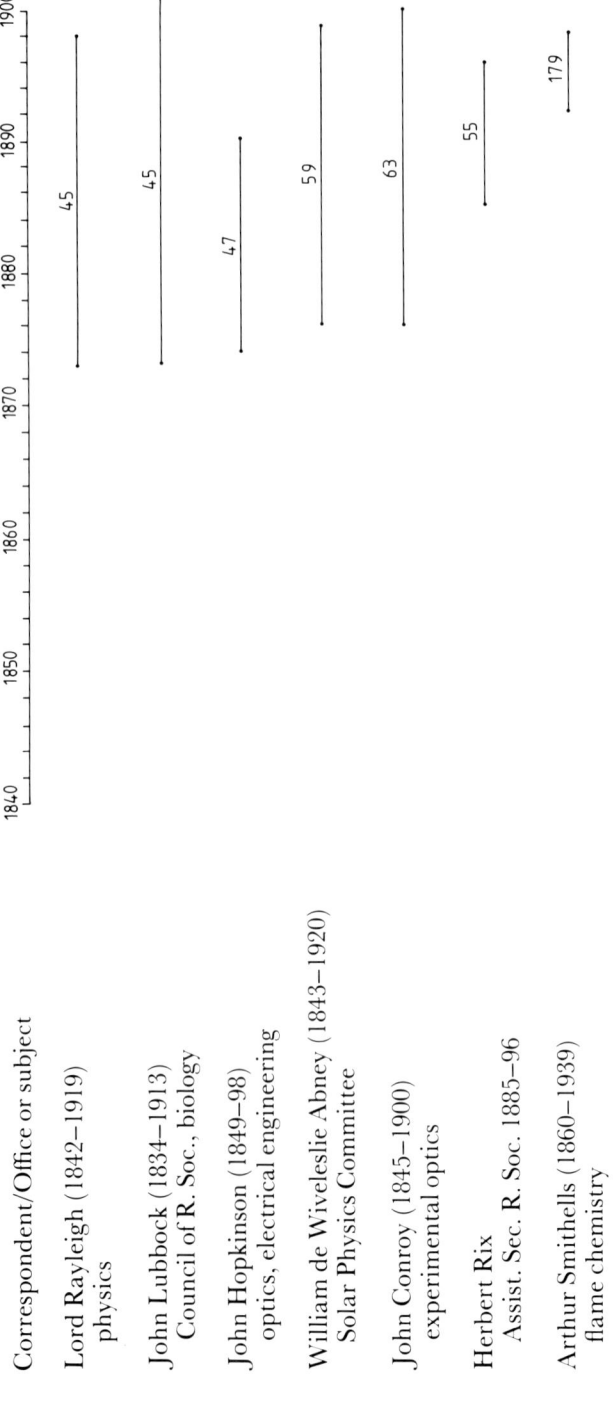

Correspondent/Office or subject

Lord Rayleigh (1842–1919)
   physics

John Lubbock (1834–1913)
   Council of R. Soc., biology

John Hopkinson (1849–98)
   optics, electrical engineering

William de Wiveleslie Abney (1843–1920)
   Solar Physics Committee

John Conroy (1845–1900)
   experimental optics

Herbert Rix
   Assist. Sec. R. Soc. 1885–96

Arthur Smithells (1860–1939)
   flame chemistry

William Vernon Harcourt, chief founder of the British Association for the Advancement of Science, had studied chemistry at Oxford and had devoted his research to investigating the effect of sustained heat on organic and inorganic materials. In 1844 he reported to the British Association on a furnace he had constructed for producing samples of glass with various chemical compositions.[12] He wanted to discover correlations between glass's chemical composition and its optical properties. Resuming his uncompleted research many years later, Harcourt asked Stokes in 1862 to test the fluorescent properties of several samples. Stokes quickly decided that their dispersive properties were more interesting, and there followed a decade-long collaboration, with Harcourt producing prisms of various chemical compositions and Stokes providing expertise in optical theory. The goal was to combine glasses of different dispersive powers into an achromatic lens. Speaking of their collaboration shortly after Harcourt's death in 1871, Stokes said ' I may certainly say for myself, and I think it will not be deemed at all derogatory to the memory of my esteemed friend and fellow-labourer if I say of him, that I do not think either of us working singly could have obtained the results we arrived at by working together.'[13] Stokes pursued his and Harcourt's research in a lecture to the Photographic Society in 1873, in reports to the British Association in 1874 and 1875, and in a paper for the Royal Society's *Proceedings* in 1878.[14] At the 1874 British Association meeting, he was able to exhibit an achromatic telescope using Harcourt's glass—the telescope being constructed by another of Stokes's correspondents, Howard Grubb.

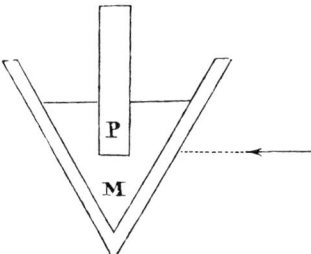

Figure 8.2 From Baden Powell 1848 'On a New Case of the Interference of Light' *Phil. Trans.* 214.

Sir John Conroy studied chemistry at Oxford with Harcourt's nephew A G Vernon Harcourt, who in 1876 communicated Conroy's first two papers on optical experiments with iodine to the Royal Society, thus commencing the Stokes–Conroy correspondence. In one long letter Stokes explained the significance of Conroy's iodine experiments to him.[15] Later, with Stokes as mentor, Conroy turned to experiments on the reflection of light from metals, a subject on which Stokes had read a paper to the British Association in 1876.[16] In particular, Stokes explained to Conroy the proper mathematical method for correcting his experimental results[17] and identified the source of discrepancy between Conroy's results and optical theory.[18] He felt it necessary also to explain to Conroy a basic aspect of physical theory: 'There is no doubt that in absorption light is *spent* in producing molecular motion. In interference light modifies light and is not *spent* at all; the distribution of illumination is altered but the sum total is not.'[19] Next, at Stokes's suggestion, Conroy undertook experiments 'On the Polarization of Light by Reflection from the Surface of a Crystal of Iceland Spar'. Such experiments, Stokes told Conroy, had been performed by David Brewster but 'had never been published in detail, and had not been repeated by anyone else'.[20] Once again, Stokes even helped Conroy manage the experimental data: 'Professor Stokes pointed out to me that the experimental results which had been obtained were well suited for reduction by means of harmonic analysis, and not only explained the method but himself reduced the first set of observations made with a cleavage face in water.'[21]

The intense and well preserved correspondence between Stokes and Arthur Smithells in the 1890s has been investigated in detail by DeKosky.[22] Though Smithells took Thomson's natural philosophy course at Glasgow, it was John Ferguson's chemistry course there which captured his career. He went on to study chemistry at Owens College, Manchester, and in Germany before becoming professor of chemistry in Leeds in 1885. Responding to a reference to himself in a letter from Stokes to the chemist, Henry Armstrong, published in the *Proceedings of the Chemical Society*, Smithells wrote Stokes 'to submit to him my views on the genesis of spectra'.[23] According to Smithells, 'I was at that time but poorly versed in the literature of spectrum analysis, but I had got the idea that gaseous spectra were much more likely to be due to disturbances occurring during the formation and decomposition of molecules than to purely thermal actions. . . .'[24] DeKosky's article explores Stokes's eventually successful efforts to persuade Smithells of the validity of the thermal explanation as opposed to the chemical. Stokes argued, for example, that a chemical reaction could release enough energy to cause a gas to glow, but that the chemical reaction itself was not the cause of the glow and that heat alone, in the absence of chemical reactions, should also cause a gas to glow. Experiments suggested by Stokes and performed by Smithells supported Stokes's side of the issue. Stokes took so much interest in this and other aspects of Smithells's research that Smithells 'found it difficult to keep up my end of the correspondence'.[25]

Robinson, Lockyer, Powell, Harcourt, Conroy and Smithells thus illustrate that prominent theme in Stokes's correspondence—his work 'in concert' with experimental scientists, often ones residing towards the chemical end of the spectrum of knowledge in the physical sciences. Stokes explained how to deal with experimental results, helped with instrumentation, suggested experiments, and explained and argued theoretical points. In all these respects, Stokes's correspondence with William Crookes was cut from the same pattern. As Brock has observed, Crookes's ignorance of mathematics 'was often masked in later years through his friendship with Stokes, who privately solved many mathematical problems in physics for him'.[26] As Oliver Lodge observed to Kelvin, 'Crookes' theoretical ideas were wild until Stokes set them straight.'[27] Indeed, given Stokes's contributions to Crookes's work, the significance of Crookes's experimental researches, and the duration of their cooperation, Stokes's correspondence with Crookes is his most important collaboration with an experimental scientist.

Stokes, Crookes and the new physics

William Crookes (1832–1919) was a man of diverse accomplishment. He edited various scientific journals, most notably *The Chemical News* and *The Quarterly Journal of Science*. His experimental work in the early 1870s on psychic phenomena was highly influential on the psychic-research movement in late-Victorian Britain. Rejected for publication by the Royal Society, his psychic research appeared largely in his own *Quarterly Journal of Science*. Having studied at the Royal College of Chemistry and served in the early 1850s as personal assistant to its director, Crookes could always rely on chemistry as his principal area of strength in the sciences. However, influenced by Faraday and the electrician, Charles Wheatstone, Crookes turned to physical aspects of chemistry for much of his research.[28]

Work in spectroscopy led to his first major accomplishment, the discovery of thallium in 1861. Trying to determine the atomic weight of thallium to the highest possible accuracy, Crookes noticed a small but unexpected phenomenon in high vacua, which he went on to investigate with devices like that in figure 8.3. When suspended in a vacuum and radiated with light or radiant heat, the horizontal 'index', *fg*, would oscillate and even rotate. Such experiments led to a series of papers published in the 1870s in the Royal Society's *Philosophical Transactions*. Midway through the series, Crookes introduced his radiometer (figure 8.4), a variation on the device in figure 8.3, which would rotate when exposed to light or radiant heat. Enormously fascinating, the radiometer, J J Thomson recalled, was 'the subject which was exciting most interest' when he began research in the 1870s.[29] Crookes initially thought that direct radiation pressure drove the radiometer, but the explanation was soon given in terms of the kinetic theory of gases, first by G Johnstone Stoney and then, with more sophistication, by Maxwell.

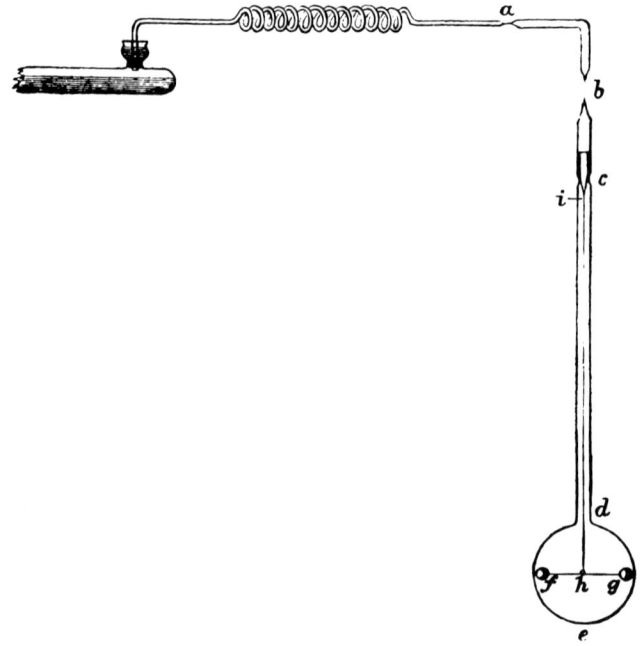

Figure 8.3 From W Crookes 1875 'On Repulsion Resulting from Radiation—Part II' *Phil. Trans.* 521. The bulb containing the horizontal index is shown connected to the air pump. '*fg* is a very fine stem of glass, drawn from glass tubing, and having a small loop (*h*) in the middle. At each end of the stem is a ball or disk, made of pith, cork, ivory, metal, or other substance.'

Adopting the basic kinetic-theory idea and introducing his concept of a 'fourth state of matter', Crookes turned in late 1878 to the phenomena of the electrical discharge in gases in a vacuum tube to confirm his views of the radiometer. He thought the same physical process occurred in the dark space around the cathode (see figure 8.5) in these electrical phenomena as occurred in the radiometer—a stream of gas molecules rebounded in parallel paths (the fourth state) away from an excited surface, either the electrified cathode or the heated radiometer disc. Highly successful, this work, for which he is most famous, persuaded British scientists to accept the 'molecular stream' theory of cathode rays. The dark space was the mean free path of the stream of electrically charged molecules, and when they finally collided with other molecules of the tube's residual gas, they gave rise to the 'negative glow'. With a sufficiently high vacuum, the dark space filled

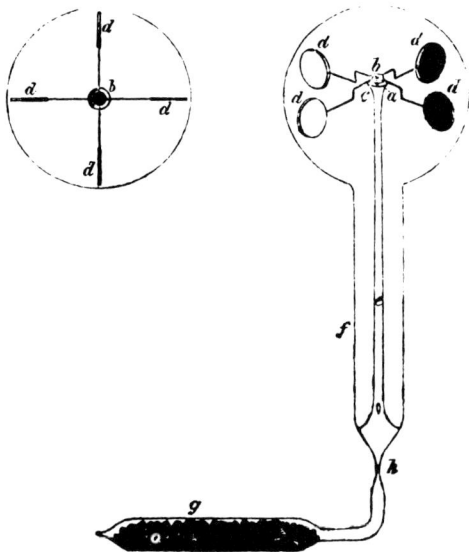

Figure 8.4 From W Crookes 1876 'On Repulsion Resulting from Radiation—Parts III & IV' *Phil. Trans.* 339. The radiometer 'consists of four arms, of very fine glass, passing horizontally through pieces of pith (*b*), and afterwards bent twice at right angles, as shown in the figure. Through the centre of the pieces of pith (*b*) is passed vertically the point of a very fine sewing-needle (*a*), which rests in a glass cup (*c*) blown on to the end of the glass tube (*e*). At the end of each glass arm is fastened a thin disk of pith, white on one side and lampblacked on the other, the black surfaces of all the disks facing the same way.'

the tube, the phosphorescence of the tube's walls being the only evidence of the stream's presence. Crookes's experiments graphically showed that the stream was deflected by a magnetic field and, when interrupted by a solid object, would cast shadows on the wall of the tube. (See figure 8.6, for example.) There followed one of the most interesting features of late-nineteenth-century physics, the British–German disagreement over the nature of cathode rays—the British following Crookes, the Germans generally thinking that cathode rays must be ethereal waves of some kind. It was thus the Briton, J J Thomson, who determined in 1897 that the supposed stream of molecules was actually a stream of electrons.

In the 1880s and into the 1890s Crookes utilised cathode rays to cause various minerals to phosphoresce, allowing him to study the rare-earth elements, which he did not think were elemental, and to make an elaborate speculation on the evolution of these elements. At the end of the century he entered the new field of radioactivity, making contributions of some importance. His theory of the source of the energy of radioactivity was one of those discussed at the time. He chemically isolated a radioactive substance which he called Uranium-X. He invented an instrument, the spinthariscope, which aided in the detection of alpha particles.

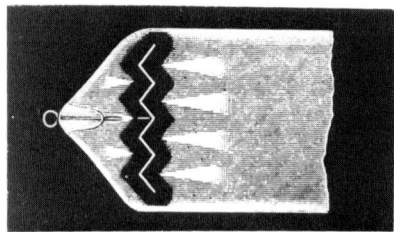

Figure 8.5 From W Crookes 1879 'Contributions to Molecular Physics in High Vacua' *Phil. Trans.* 643. In this experiment the pressure of the gas is great enough to keep the dark space from expanding to fill the tube. Suggested by Maxwell, the experiment supported Crookes's idea that the molecular stream was projected normal to the surface of the negative pole.

This section of the chapter concentrates on Crookes's work on cathode rays and radioactivity, 'the new physics'. This label, though appropriate for radioactivity, may seem misleading for cathode rays. Research there neither required nor immediately led to new physical principles. However, in a different sense than for radioactivity, the label is suitable. The enormous importance of cathode-ray research was recognised at the time and was not merely proclaimed with post-1897 hindsight. For example, Stokes's referee's report on Crookes's first paper on cathode rays stated: 'I consider this a most remarkable paper, throwing a flood of light on the behaviour of the ultimate molecules of matter in the gaseous state. . . . I look on these investigations as forming a great step in the progress of physical science.'[30] We have already cited Kelvin's 1893 opinion, which concluded praise of various experimenters, especially Crookes, with the statement: 'If a first step towards understanding the relations between ether and ponderable matter is to be made, it seems to me that the most hopeful foundation for it is knowledge derived from experiment on electricity in high vacuum.'[31] Hence, experimental investigation of cathode rays was providing fundamentally *new* glimpses of nature's hidden realm, to the delight of those cautious realists, Kelvin and Stokes.

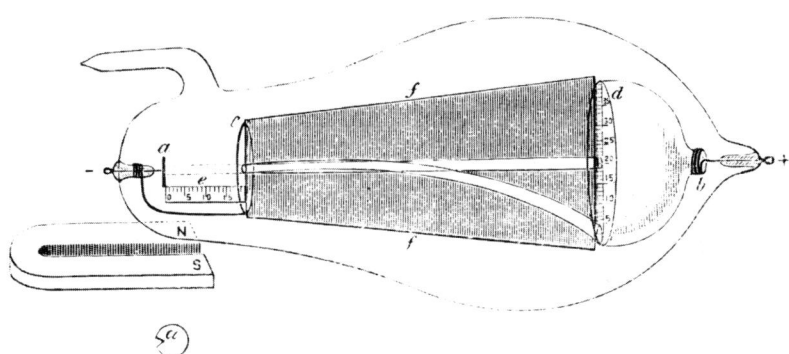

Figure 8.6 From W Crookes 1879 'On the Illumination of Lines of Molecular Pressure, and the Trajectory of Molecules' *Phil. Trans.* 157. The molecular stream coming from the negative pole *a* passes through a hole in the mica screen *c* and falls on the flat plate of German glass *d*. 'The notch in the pole is to enable me to see if the spot of light projected on *d* is an image of the pole *a*, or of the hole in *c*.' The curved ray is, of course, the one being deflected by a magnetic field, though in the actual experiment one would observe only the spot of light, not the ray itself. Crookes usually found the image to have no notch, a result he took to support the theory that the rays consisted of streams of molecules, not waves of light. In the latter case, the hole in *c* would have projected on *d* an inverted image of the notched pole. (*Ibid.* p. 161.)

Stokes's correspondence with Crookes began in the mid-1850s when Crookes asked Stokes to identify certain spectral lines for him.[32] In 1862 Crookes asked Stokes for a testimonial, having 'so frequently been in communication with you on scientific subjects'.[33] Their period as regular correspondents, however, began in June 1871 in connection with Crookes's submission of his paper on psychic force to the Royal Society. In a highly polemical paper in his own journal in October 1871, Crookes published two of Stokes's letters to him plus a short report that Stokes had been asked to write on a psychic paper submitted by Crookes to the British Association.[34] As secretary of the Royal Society and as *ad hoc* advisor to section A of the British Association, Stokes was the one person most obviously preventing Crookes from presenting his findings to the scientific community. Crookes's October 1871 paper did not adopt a friendly tone towards Stokes. Uneasy exchanges continued through the spring of 1872, but by autumn of 1873 Crookes was presenting other research to the Royal Society and was grateful for Stokes's criticism, even though, as he admitted to Stokes, he had been reluctant to ask for it directly: 'I had some thoughts of submitting the proof to you before sending it in and asking you to kindly look it over to see if I were correct; but just before, we had had a correspondence respecting a paper on some other subject, and I had an idea (wrongly I now believe), that you would rather not be bothered by looking over any more of my papers.'[35] Thus reconciled, the two actively corresponded on Crookes's experimental researches over the next quarter of a century.

Fortunately, in different ways, a large part of the Stokes–Crookes correspondence has survived. As usual, Stokes appears to have kept virtually all of his letters from Crookes. There are 156 in the Stokes Collection. Crookes also 'kept religiously every letter I have received from Stokes',[36] but they have all disappeared, evidently some time after E E Fournier D'Albe published his biography of Crookes in 1924. Fournier D'Albe did not specifically mention Stokes's letters, but they were undoubtedly among the some 40 000 letters and documents which, he said, 'came into my hands' for writing the book.[37] However, Crookes had sent a *printed* version of many of Stokes's letters to Joseph Larmor, who included 119 in his *Memoir and Scientific Correspondence of the Late Sir George Gabriel Stokes*. In addition, the *Memoir* includes 27 Crookes-to-Stokes letters not in the Stokes Collection, and the Stokes Collection contains four copies of Stokes-to-Crookes letters not in the *Memoir*. Finally, the Royal Society holds two letters from Crookes to Stokes, and one of Stokes's letters that Crookes published in 1871 does not survive elsewhere. Altogether, counting originals, copies and printed versions, 184 letters written by Crookes and 126 by Stokes survive. Though the originals of almost all the letters used by Larmor are now missing, fortunately for this chapter he regarded the late 1870s and the period around 1900 as important stages in the cooperation between Stokes and Crookes. Hence, the some sixty missing letters date mainly from the early 1870s and the 1880s.

At least two aspects of the Stokes–Crookes collaboration were clearly established in the 1870s, well before Crookes began his work on cathode rays. First, Crookes readily acknowledged Stokes's vast superiority in mathematics. There were, for example, enough mathematical errors in Crookes's paper on the atomic weight of thallium that its publication was delayed. 'Not feeling strong enough mathematically to deal fully with the subject',[38] Crookes explained, he had relied on a former assistant to Augustus DeMorgan. After Stokes's corrections, Crookes told him that 'authors who are not strong in mathematics owe you an everlasting debt of gratitude for your pains-taking revision'.[39] In the same vein, in much of the correspondence of the mid- to late 1870s, Stokes carried out the mathematical analysis of Crookes's results, explaining to Crookes how to take the readings, how to deal with 'methodical' and 'casual' errors, and what the results meant.

Second, Stokes was the arbiter of physical theory, though that is not to say that Crookes always agreed totally with him or had no ideas of his own. However, Stokes's dominant role can be seen repeatedly. He corrected Crookes's use of *weight* and *mass*, for example, and Crookes's reply disclosed a somewhat unsure grasp of the point: 'It is a bit of carelessness however into which great authorities fall occasionally, and although weight usually refers to mass, strictly speaking it is I think a measure of *force.*'[40] Crookes wanted to entitle the first of his series of radiometer papers 'On Attraction and Repulsion Accompanying Radiation' rather than, as Stokes preferred, 'On Attraction and Repulsion Resulting from Radiation'. He argued to Stokes, rather curiously:

> Now Radiation is an abstract term. It relates to no body or substance, but only to an attribute of something. To talk of attraction and repulsion resulting from an attribute is therefore incorrect. Were I to say *accompanying* Radiation all the metaphysical difficulty would disappear, because attraction and repulsion and radiation would be used together abstractedly. Attraction may accompany, but cannot result from, an abstract quality.[41]

Stokes's answer is lost, but he had his way. When Crookes came round to Stokes's acceptance of the kinetic-theory explanation of the radiometer, he relied on Stokes's expression of the theory.[42] Probably most important, it was Stokes who defined the importance of Crookes's work in the 1870s to current physical theory. Crookes's high vacua, Stokes pointed out, provided a means of confirming Maxwell's con-clusion that the viscosity of a gas was independent of its density.[43] Even more interesting, Stokes thought, Crookes had a way of investi-gating the respective roles of radiation and conduction in the transfer of heat—an investigation that, among other things, might help test William Huggins's recent conclusion that comets were incandescent.[44] Thus guided by Stokes, Crookes eventually published papers on both the conduction of heat in high vacua and the viscosity of gases, the latter paper being followed in the *Philosophical Transactions* by Stokes's own discussion of the topic.[45]

Completion of these investigations was sidetracked in the autumn of 1878. In early October, Crookes wrote to Stokes that 'I should like to shew you some curious phenomena which I have been working at lately. . . . I have unearthed some new facts in connection with this subject which are of absorbing interest, and make me begrudge every minute I give to other subjects. . . . I hope to show you the molecules at work.'[46] By mid-October Stokes had visited Crookes's home to see 'your most interesting experiments'.[47] In early December Crookes presented his results to the Royal Society in what was his first paper on cathode rays.[48] Still the dominant theorist, Stokes corresponded with Crookes on the subject from October 1878 to August 1879.

Crookes's most staggering claim was that he had found evidence supporting the corpuscular theory of light. His basic idea was that light was corpuscular both in his laboratory vacuum and in the similar emptiness of space, but otherwise was undulatory. In November he put these and other ideas in a letter to Stokes:

> May it not be that the interplanetary space is filled with this molecular matter? The sun may then act like the negative pole of the coil and drive the molecules in all directions from it. An individual molecule need not actually pass from the sun to the earth, but a chain of molecular impacts may carry the force along in a straight line. When the molecules meet the earth's atmosphere bombardment takes place with production of heat & light, just as when the molecules hit the glass. The blue colour of the sky is the colour of phosphorescence. On high mountains there is less bombardment of the atmosphere and more of the solid substance. The idea is expressed very crudely and may not be worth following out, but the more I think it over the more it assumes consistency, and it does appear that what Newton thought was true is not far from my suggestion, and if so it cannot be very absurd. We thus should have *emissive* light (?) in space and undulatory light in connection with denser matter.[49]

Stokes's response does not survive, but in the first version of Crookes's paper, Crookes still regarded the molecular stream from the negative pole as 'rays of molecular light'. In concluding the abstract of his paper for the *Proceedings*, he claimed that 'the author . . . is as well entitled to speak of a ray of molecular or emissive light when its presence is detected only by the light evolved when it falls on a suitable screen, as he is to speak of a sunbeam in a darkened room as a ray of vibratory or ordinary light when its presence is to be seen only by interposing an opaque body in its path. . . . The light by which the eye then sees the screen is, of course, undulatory.'[50] Stokes discounted Crookes's idea both in his short report to the Society's Committee on Papers and, evidently, in a more detailed letter to Crookes.[51] Crookes resisted, declaring: 'I don't see much difference between the line of emitted molecules in these tubes, and Newton's line of corpuscles.'[52] Stokes explained: 'The difference between the ray of emitted molecules and what Newton *supposed* to be a line of corpuscles is that the latter is capable of exciting the sensation of light when received directly into the eye. But nobody now-a-days believes that what Newton supposed to be a line of corpuscles is really such.'[53] The upshot was that Crookes's giant claim made in his abstract shrank in the later, full version of the paper to the lone, enigmatic statement that his experiments disclosed a world 'where the corpuscular theory of light may be true, and where light does not always move in straight lines'.[54]

Stokes was much more interested in the cause of the tube's fluorescence—radiation or molecular impact? The sharp shadows cast in Crookes's tubes convinced Stokes that Crookes was correct in envisaging streams of molecules, not waves of light, emanating from the negative pole. Additionally, the pole was not hot enough to be incandescent and produce ordinary light.[55] Stokes thought Crookes was probably right that actual molecular impacts produced the fluorescence. Yet, since that would have been an entirely new phenomenon (fluorescence previously being produced only by ethereal waves), Stokes thought it 'desirable to put the explanation to whatever tests one could think of'.[56] It was possible, for example, that the molecules became incandescent when they reached the far end of the tube, producing radiation which, in turn, was the immediate cause of the tube's fluorescence. Stokes reasoned that molecular impacts would cause only the inner surface of the tube's glass to fluoresce whereas radiation would cause fluorescence within the glass. Accordingly, Stokes proposed the experiment illustrated by Crookes in figure 8.7, in which *a* was a plate of uranium glass and *b* and *c* were thin films of uranium glass and German glass, respectively. Stokes explained the possible results to Crookes:

> Now, if the fluorescence be due to actual impact, the fluorescent light ought to be as strong on the very thin glass as on the thick piece of the same kind, and, moreover, the thin pieces ought to cast a perfect shadow. But if it be due to radiation we should expect that the full amount of fluorescence would not be developed on such excessively thin pieces, and, moreover, that some of the rays would be able to get through and affect what lay beneath.[57]

Crookes reported that the results favoured molecular impacts.[58]

This experimental result leads naturally into the sharp division between British and German physicists. Though it is possible dispassionately and objectively to compare the arguments and evidence for the two views, judiciously recording their respective strengths and weaknesses,[59] the striking *national* split over the issue suggests that that it not entirely what happened historically. Three features of the Stokes–Crookes correspondence may help explain how the British view developed.

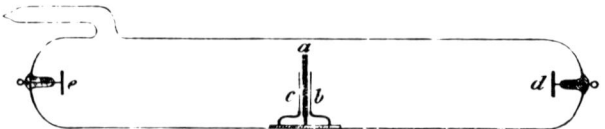

Figure 8.7 From W Crookes 1879 'On the Illumination of Lines of Molecular Pressure, and the Trajectory of Molecules' *Phil. Trans.* 150.

First, the Stokes–Crookes correspondence contains no hint of a German theory of the phenomena. They considered the research of others on subjects related to the phenomena, but ignored Crookes's German precursors, Plücker, Hittorf and Goldstein, even though much of the German research existed in English translation. Apparently, the research was either unknown or not perceived to set forth a particular theory of the phenomena. The former was the case with Goldstein's work, which Crookes presented both in a long footnote in his own paper and in summary form in *The Chemical News*.[60] In each instance, Crookes said that his own research had been done in ignorance of Goldstein's. However, the early papers by Goldstein, Hittorf and Plücker did not, in fact, present an ether-wave theory of the phenomena. Crookes was not, for example, acknowledging ignorance of an ether-wave theory of Goldstein's but ignorance of Goldstein's experimental results, some of which Crookes had duplicated. Even though the Germans' papers spoke of the 'negative light' being propagated in rays from the cathode and as causing the tube's phosphorescence, such passages seem to be discussions of the negative *glow* itself, not arguments that *ether* waves *caused* the glow and the phosphorescence.[61] Whatever German physicists might have said to themselves in conversation or correspondence, their publications before 1878 did not argue for an ether-wave theory of cathode rays. Significantly, the 1877 edition of a leading textbook produced by one of William Thomson's students—although it discussed electrical discharge in rarefied gases and displayed coloured pictures of the phenomena as a frontispiece—did not mention any theory of cathode rays.[62]

Second, Stokes gave substantial, if not unequivocal, support to Crookes's idea of a fourth state of matter, the concept which underlay the transition from radiometer to cathode rays. Shortly after publication of Stoney's articles on the radiometer in early 1876, Stokes emphasised (as Stoney had not) that Crookes's vacua were great enough 'that the mean length of path of the molecules between their collisions is no longer very small compared with the dimensions of the apparatus'.[63] Crookes incorporated the idea in papers sent to the Royal Society in June and November 1876, in the latter introducing his notion of a fourth state of matter.[64] Stokes agreed that Crookes had attained a non-gaseous state, only disagreeing that this was a *fourth* state: 'I should say that it was a state in which properties of matter which exist even in the gaseous state are shown *directly*. . . .'[65] Thus, in contrast to Crookes's attempt to resurrect the corpuscular theory of light, his idea of a fourth state of matter may well have been initially suggested to him by Stokes and, at any rate, was endorsed in its essentials by Stokes. Stokes's theoretical judgement reinforced Crookes's explanation of cathode rays.

Third, between them, Stokes and Crookes had devised and performed a direct experimental test for ethereal radiation as the cause of the tube's phosphorescence. Over a decade later J J Thomson declared that this very experiment not only contradicted the theory that cathode rays 'are a kind of ethereal vibration', but that, indeed, 'the sharpness and blackness of these shadows are by far the strongest arguments in support of the impact theory of the phosphorescence'.[66]

Let us summarise the events of the 1870s. Prior to Crookes's work, German workers in the area had not proposed a theory of the *cause* of the negative glow. In 1871, however, a Briton had. Finding that a magnetic field could curve the negative glow into the shape of an arch and that a strip of talc inserted into the arch would cast a shadow, C F Varley briefly suggested that the glow 'is composed of *attenuated particles of matter projected* from the negative pole by electricity in all directions'.[67] In the mid-1870s Crookes's radiometer experiments aroused enormous interest and controversy, and led directly to his molecular-stream theory of the cause of the negative glow and the tube's phosphorescence. Attention paid to Crookes's radiometer research ensured a ready audience for his cathode-ray work. One can thus imagine a kind of shared experience among British physicists as Crookes and Stokes, *in the absence of competing theories*, worked in concert to develop a convincing viewpoint, which even contained a direct experimental refutation of any ether-wave theory of the phenomena. Because of their connections with Varley, Crookes and Stokes should certainly have been aware of his previous work, and perhaps others were as well.[68] Insofar as Varley's theory was known, the Crookes–Stokes view would have seemed a natural extension and refinement of its main idea. Consequently, though there existed differences over detail, the stream-of-particles theory became the standard British view of cathode rays.

Just how well established it became can be seen in an exchange of letters between Kelvin and Stokes in 1896. Kelvin observed that the German physicist, Philipp Lenard, had been 'punished' for following the German 'perversity': 'The absurd supposition of a Cathode undulatory ray and no torrent of molecules lost the discovery of the X-rays to Lenard.'[69] Stokes replied:

> I was so satisfied that the ray-like things that come from the kathode, and which are acted on by a magnet, are streams of molecules, or perhaps atoms, that when I came accidentally across the expression "Kathodostrahlen" [*sic*], I supposed they were called rays only in a metaphorical sense, as a convenient term. On mentioning this some little time ago to [J J] Thomson, he told me that there were several who took them for actual rays of some kind. On subsequently reading some German papers I found that the writers were so strongly of that opinion that they rather I think looked down on those who thought otherwise.[70]

Rather than accept Lenard's interpretation of his chief experiment in support of an ether-wave theory, Stokes proposed a mechanical analogy to explain the result on Crookes's theory. In Lenard's experiment, as Stokes said, cathode rays 'passed, or seemed to pass, through an aluminium window into a vacuum as nearly perfect as he could make it'.[71] Because an ether wave, but not a material molecule or atom, could pass through a thin metal foil, Lenard's finding was supposed to support the German theory. Stokes, however, wrote: 'My notion is that it is something like the way in which, in a row of ivory balls in contact, if the first be struck the last flies off; not that I take so gross a merely mechanical notion as that.'[72] Responding, Kelvin was amazed at how the Germans 'have insisted with something like strong partisan spirit upon a ray of undulatory light from the Cathode and have blindly refused to accept Varley's conclusion of a torrent of molecules, corroborated as it is by Crookes'.[73] Therefore, it was evidently the *1890s* before Stokes even became aware of the German ether-wave theory, by which time developments had persuaded him and Kelvin that Crookes's theory was obviously correct.

In the autumn of 1896 Stokes repeated his support to Crookes himself: 'According to my knowledge so far, I am very strongly disposed to believe that with respect to the nature of the so-called 'Kathodenstrahlen' you are right, and Wiedemann, Lenard and Röntgen are wrong.'[74] Moreover, during this very time, 1896–7, Stokes was trying to form an integrated theory of light, Röntgen rays, Becquerel rays and cathode rays, with Crookes's theory as a starting point for the enterprise. As he wrote to Crookes in 1897: 'I have come to pretty definite views regarding the nature of the Röntgen rays. . . . The whole thing is based on the view that the 'Kathodenstrahlen' are not rays at all, but streams of molecules.'[75]

We saw in the last chapter that by February 1896 Kelvin and Stokes had agreed that Röntgen rays must be high-frequency transverse waves. Only a month later, Stokes broached to Kelvin an alternative theory, according to which the sudden halt of a cathode-ray molecule in the wall of the tube gave rise to an x-ray pulse.

> My mind has rather turned from the idea of excessively rapid vibrations excited within molecules . . . which communicate themselves to the ether, to the idea of a vast succession of independent pulses in the ether, each charged molecule producing such a pulse when it gets to the wall. Contrasted with the other idea, this would be like the sound arising from the hedge fire of a regiment as contrasted with the ring of rapidly (I mean as to succession, not velocity) struck bells.[76]

Though still supporting the idea of transverse waves in the spring of 1896,[77] in a series of lectures over the next 18 months Stokes gave increasing attention to the 'hedge fire' analogy, fully endorsing this pulse theory in his Wilde lecture in Manchester in July 1897.[78] Tied theoretically to the molecular-stream view of cathode rays and linked experimentally to the absence of reflection, refraction, diffraction, and interference for Röntgen rays, Stokes's pulse theory, as Wheaton has explained, became the prevailing view of x-rays during the first decade of the new century.[79]

Related to Stokes's theory of Röntgen rays was his 'wagtail' conjecture regarding Becquerel rays. After Becquerel's discovery in early 1896, it was natural to compare radiations with one another. Becquerel's experiments showed that, like Röntgen rays, Becquerel rays could penetrate opaque materials to expose photographic plates and that, unlike Röntgen rays, they were polarised and could be refracted. Accordingly, in his Rede lecture at Cambridge in the summer of 1896, J J Thomson declared that Becquerel rays were 'undoubtedly' a form of light and that they formed 'a link between the Röntgen rays and ordinary light'.[80] In agreement with Thomson's views, Stokes's wagtail conjecture went further and tried to get at the origin of Becquerel rays. Ordinary waves of light arose from regular, sustained vibrations of or within ordinary molecules. Röntgen-ray pulses arose from the irregular, sudden stops of cathode-ray molecules. The origin of Becquerel rays, therefore, ought to involve a mechanism in between. In his Wilde lecture, Stokes offered a mechanical analogy:

> My conjecture is that the molecule of uranium has a structure which

may be roughly compared to a flexible chain with a small weight at the end of it. Suppose you have vibrations communicated to such a chain at the top; they travel gradually to the bottom, and near the bottom produce a disturbance which deviates more from a simple harmonic undulation. So, if a vibration is communicated to what I will call the tail of the molecule of uranium, it may give rise to a disturbance in the ether which is not of a regular periodic character.[81]

Growing out of Stokes's discussions—quite successful discussions—of cathode rays and Röntgen rays, his wagtail theory constituted an essential part of the background to the final phase of his correspondence with Crookes.

First, they tried to decide experimentally between Stokes's wagtail theory and Crookes's 'kick' theory. Proposed the year after Stokes's Wilde lecture in his presidential address to the British Association, Crookes's theory compared a radioactive atom to Maxwell's demon. Realising that radioactive elements required neither optical nor electrical excitation in order to emit their rays, Crookes sought another source of energy and found it in molecules of air. Like a Maxwellian demon, Crookes suggested, radioactive atoms could 'have a structure that enables them to throw off the slow moving molecules of the atmosphere, while the quick moving molecules, smashing on to the surface, have their energy reduced and that of the target correspondingly increased'.[82] Though Stokes's and Crookes's hypotheses emphasised different aspects of radioactivity—Stokes's the process producing Becquerel rays, Crookes's the source of the energy—they nevertheless sought to choose between them. To do so, they decided to compare the luminosity of radium in air and in a vacuum. Stokes explained that his theory predicted no difference but that 'according to the idea of kicks by molecules of air I should expect the luminosity to be, not perhaps destroyed, since no vacuum we can make is perfect, but at any rate greatly reduced'.[83] Performing the experiment, Crookes first detected no difference,[84] a result, Stokes wrote, 'in accord with what, in speaking to you, I called the 'wagtail' theory of the Becquerel rays'.[85] A few days later, however, Crookes decided there was a 'slight difference' in favour of the radium in air. 'This gives the bombardment hypothesis an advantage over the wagtail one', he wrote.[86] Disagreeing, evidently because the difference was too small, Stokes proposed an analogy to his wagtail analogy in order to suggest why the presence of air might slightly increase the intensity of Becquerel rays:

To speak metaphorically, if the dog wags his tail so as at each wag to strike against a half opened door, the agitation he produces in the air may be greater than if the wagging tail did not come against anything but air. For air read ether; for the half opened door read a molecule of air; for the dog's tail read a sort of termination of a molecular structure, agitations propagated into which become more smart, in a manner analogous to the agitation produced about the lower end of a chain suspended from one end, produced by an agitation at the upper end, which travels down the chain.[87]

Implicit in the correspondence was Stokes's apparent assumption that the activity of radium would not be affected by external circumstances. That was not his original view. In 1899, the year before his correspondence with Crookes, Stokes wrote to Becquerel that his theoretical ideas led him to think that a substance's activity 'would depend very materially upon its temperature'.[88] He supposed that the molecular structure of a radioactive substance could experience vibrations which were 'excited and sustained by the ethereal vibrations which emanate from the various bodies in the environment'.[89] Stokes suggested an experiment to test his idea, but Becquerel reported that some of his unpublished work showed there was no change over a temperature range of 120 °C.[90] Thus, the next year it would seem, in Stokes's mind the wagtail theory excluded external influences on radioactive substances.

Their second main topic—the nature of Becquerel radiation itself—primarily illustrates the propensity to follow one's own reasoning as far as possible. By 1900 experiments had indicated that Becquerel radiation was not homogeneous. Stokes referred, for example, to R J Strutt's experiments in which a magnet deflected parts of the radiation from radium in different amounts, separating them from one another.[91] But what *were* these different radiations?

Stokes's main conjecture in the area of radioactivity—his wagtail theory—postulated that a particular kind of molecular vibration produced a particular kind of ether wave. Accordingly, he interpreted Strutt's results as having divided Becquerel radiation into non-deflectable 'rays' in the ether and deflectable 'jets' of material particles. 'Whether the propelled matter came from the substance of the radium or from the air (or other gas) in contact with it, I leave an open question, but I think the latter more probable.'[92] The latter also, of course, followed naturally from the wagtail theory, according to which radium molecules produced *ethereal* vibrations.

The overriding theory of Crookes's career was the fourth state of matter, which he took to be confirmed by J J Thomson's discovery of the electron. At the end of 1900, therefore, he seemed ready to explain not only cathode rays and the different Becquerel rays but Röntgen's as well in terms of 'Thomson's ultra-atomic particles'. Referring to his experiments 20 years earlier which showed that the magnetic deflection of cathode rays was inversely proportional to their velocity, Crookes

Stokes in 1892 at the age of 73. Photograph by Mrs F W H Myers. From Stokes 1880–1905 *Mathematical and Physical Papers* ed J Larmor vols IV and V (Cambridge: Cambridge University Press) vol. V frontispiece.

Kelvin in 1897 at the age of 73. Photograph by Annan, Glasgow. From Silvanus P Thompson 1910 *The Life of William Thomson, Baron Kelvin of Largs* 2 vols (London: Macmillan) vol. I frontispiece.

wrote to Stokes, 'I consider that the magnetic deflection of the [Becquerel] rays is simply a question of degree.' The non-deflection of some Becquerel rays was 'proof that they are projected with too great a velocity to be affected by the magnet, or perhaps they carry a too small charge'. Furthermore, 'I should use similar arguments to show that the Röntgen rays were projections and not (directly) affections of the ether.'[93]

Not only was there a conceptual relationship between cathode rays and radioactivity—though Stokes and Crookes differed on exactly what it was—there was also a consistent relationship between Stokes and Crookes in their collaboration on the two subjects. As before, Crookes was the superb experimenter with ideas of his own but ready to try any experiment that Stokes recommended. As before, Stokes was the representative of sound physical theory. Surely, Stokes's implicit assumption that energy was stored within radioactive elements, for example, was superior to Crookes's invoking a Maxwellian demon-like process, rendered implausibly improbable by the second law of thermodynamics. Moreover, unlike Crookes's suggestion that Röntgen rays were electrons, Stokes's pulse theory of Röntgen rays fits into an integrated theoretical understanding of the causes not only of Röntgen rays but of Becquerel rays and ordinary light as well.

A Victorian correspondent

Though the influence of Stokes and Crookes in the area of radioactivity was to be far less than it had been in the understanding of cathode rays, both episodes reflect the kind of partnership Stokes so often formed with experimental scientists. We can thus better appreciate the accuracy of Smithells's tribute: 'He was willing to lend his time and his talents without stint to anyone whom he deemed an honest worker, and the only thing he seemed to exact in return was that there should be no fuss or effusiveness of acknowledgement. . . . What Stokes did for his generation can hardly be estimated.'[94]

Moreover, we can now see Stokes as a kind of bastion of orthodoxy in Victorian physical science. Thrust into and sustained in the role mainly by his Royal Society secretaryship, he occupied his position because of the institutional structure of Victorian science as well as because of his own considerable abilities. He could correct the misunderstandings of a Conroy or a Crookes or challenge the concepts, like that of an aerial ether, of a profoundly knowledgeable William Thomson. As none other did, Stokes, as guide and critic, functioned as almost an official interpreter of what constituted proper physical theory.

Finally, at the close of the century, Stokes brought Victorian physical theory to bear on the new physics of x-rays and radioactivity. His theory of x-rays was highly influential. His more tentative hypothesis regarding the production of Becquerel rays, on the other hand, vanished in the enormous theoretical–experimental transformation that occurred in that field over the next few years. Still, the issues about which he and Crookes corresponded—the nature of Becquerel rays and the source of the energy of radioactivity—remained key problems. Not until the autumn of 1902, only a few months before Stokes's death, did Rutherford finally work out a stable classification of the Becquerel radiations.[95] And it was not until shortly after Stokes's death in 1903 that it became clear just how staggering were the implications of Stokes's apparent assumption that, contrary to Crookes's ideas, radioactive energy resided *within* the radioactive substance. Indeed, it was at that point that Kelvin began challenging this 'internal' view of radioactive energy, now represented above all by the young Ernest Rutherford.

1   Stokes 1862 'Address' *Br. Assoc. Rep.* part 2 p. 1.
2   W Huggins 'Appreciation [of Stokes]' *Memoir* I 103.
3   H Grubb to J Larmor, 28 August 1905 *CUL Stokes Collection* L111.
4   Michael Faraday to James Clerk Maxwell, 25 March 1857 in L Pearce Williams (ed) 1971 *The Selected Correspondence of Michael Faraday* 2 vols (Cambridge: Cambridge University Press) II 864.
5   Michael Faraday to James Clerk Maxwell, 13 November 1857 in *ibid.* II 885.
6   Stokes to John Tyndall, 1 January 1862 *Tyndall Papers, Royal Institution.*
7   Stokes's appendix to T R Robinson 'On the Determination of the Constants of the Cup Anemometer by Experiments with a Whirling Machine' *MPP* V 95–9.
8   On the Stokes–Schunck correspondence see W V Farrar 1977 'Edward Schunck, F.R.S.: A Pioneer of Natural-Product Chemistry' *Not. Rec. R. Soc.* **31** (January) 273–96.
9   There was apparently no correspondence with Tomlinson during these years.
10 A J Meadows 1972 *Science and Controversy: A Biography of Sir Norman Lockyer* (Cambridge, MA: MIT Press) pp. 118–19.
11 Baden Powell 1848 'On a New Case of the Interference of Light' *Phil. Trans.* 213. Stokes's paper, published immediately after Powell's in the *Phil. Trans.* is 'On the Theory of Certain Bands Seen in the Spectrum' *MPP* II 14–35. Larmor discusses the Stokes–Powell collaboration in *Memoir* I 115–24.
12 William Vernon Harcourt 1844 'Report on a Gas Furnace for Experiments on Vitrifaction and Other Applications of High Heat in the Laboratory' *Br. Assoc. Rep.* 82–5.

13   Stokes 'Notice of the Researches of the Late Rev. W. Vernon Harcourt on the Conditions of Transparency in Glass and the Connection between the Chemical Composition and Optical Properties of Different Glasses' *MPP* IV 340.

14   Stokes 'On the Principles of the Chemical Correction of Object-Glasses' *MPP* IV 344–54; 'On the Construction of a Perfectly Achromatic Telescope' *MPP* IV 356–7; 'On the Optical Properties of a Titano-Silicic Glass' *MPP* IV 358–60; and 'On an Easy and at the Same Time Accurate Method of Determining the Ratio of the Dispersions of Glasses Intended for Objectives' *MPP* V 40–51. Comments by Larmor plus a few Stokes–Harcourt letters are in *Memoir* I 200–2; II 86–99.

15   Stokes to John Conroy, 1 April 1876, *Balliol College Conroy Papers;* Conroy 1876 'Absorption-Spectra of Iodine' *Proc. R. Soc.* **25** 46–51; and 'On the Polarization of Light by Crystals of Iodine' *Proc. R. Soc.* **25** 51–60.

16   Stokes 'On a Phenomenon of Metallic Reflection' *MPP* IV 361–4.

17   John Conroy 1879 'Some Experiments on Metallic Reflexion' *Proc. R. Soc.* **28** 248, 250, and John Conroy 1881 'Some Experiments on Metallic Reflexion. No. II' *Proc. R. Soc.* **31** 489.

18   John Conroy 1883 'Some Experiments on Metallic Reflection. No. III: On the Amount of Light Reflected by Metallic Surfaces' *Proc. R. Soc.* **35** 36; Stokes's note to Conroy's paper *ibid.* pp. 39–41; and Conroy 1884 'Some Experiments on Metallic Reflection. IV: On the Amount of Light Reflected by Metallic Surfaces. II' *Proc. R. Soc.* **36** 193.

19   Stokes to John Conroy, 2 June 1877 *Balliol College Conroy Papers.*

20   John Conroy 1886 'On the Polarization of Light by Reflection from the Surface of a Crystal of Iceland Spar' *Proc. R. Soc.* **40** 175.

21   *Ibid.* p. 186.

22   Robert K DeKosky 1980 'George Gabriel Stokes, Arthur Smithells and the Origin of Spectra in Flames' *Ambix* **27** 103–23.

23   Arthur Smithells to Joseph Larmor, 6 July 1905 *Memoir* I 265. Stokes 'On the Reactions Occurring in Flames' *MPP* V 235–7.

24   Smithells to Larmor, 6 July 1905.

25   *Ibid.*

26   W H Brock 1971 'Crookes, William' *Dictionary of Scientific Biography* vol. III (New York: Scribner) p. 474.

27   Oliver Lodge to Kelvin, 6 February 1896 *CUL Kelvin Collection* L82.

28   On Crookes see Brock 'Crookes, William' (note 26); E E Fournier D'Albe 1924 *The Life of Sir William Crookes* (New York: D Appleton); A E Woodruff 1966 'William Crookes and the Radiometer' *Isis* **57** 188–98; S G Brush and C W F Everitt 1969 'Maxwell, Osborne Reynolds, and the Radiometer' *Hist. Stud. Phys. Sci.* **1** 105–25; R K DeKosky 1973 'Spectroscopy and the Elements in the Late Nineteenth Century: The Work of Sir William Crookes' *Br. J. Hist. Sci.* **6** 400–23; R K DeKosky 1976 'William Crookes and the Fourth State of Matter' *Isis* **67** 36–60; R K DeKosky 1983 'William Crookes and the Quest for Absolute Vacuum in the 1870s' *Ann. Sci.* **40** 1–18; F A J L James 1981 'The Letters of William Crookes to Charles Hanson Grenville Williams 1861–2: The Detection and Isolation of Thallium' *Ambix* **28** 131–57;

and F A J L James 1984 'Of "Medals and Muddles": The Context of Discovery of Thallium: William Crookes's Early Spectro-Chemical Work' *Not. Rec. R. Soc.* **39** 65–90. For a survey of his psychical research, see J Oppenheim 1985 *The Other World: Spiritualism and Psychical Research in England, 1850–1914* (Cambridge: Cambridge University Press) pp. 338–54. I am grateful to Professor DeKosky for advice and assistance regarding Crookes's manuscripts.

29   J J Thomson 1936 *Recollections and Reflections* (London: G Bell) p. 372.

30   Stokes's referee's report on W Crookes 1879 'On the Illumination of Lines of Molecular Pressure, and the Trajectory of Molecules' *Phil. Trans.* **170** 135–64, Royal Society of London RR 8.151.

31   Kelvin 'Extract from Address of November 30, 1893' *PLA* II 557.

32   Stokes to Crookes, 2 March 1856 *Memoir* II 363–5.

33   Crookes to Stokes, 2 July 1862 *CUL Stokes Collection* C1062.

34   Crookes 1871 'Some Further Experiments on Psychic Force' *Q. J. Sci.* **1** 471–93.

35   Crookes to Stokes, 23 September 1873 *CUL Stokes Collection* C1076.

36   Crookes to Joseph Larmor, 16 October 1904 *Memoir* II 362.

37   Fournier D'Albe *Sir William Crookes* (note 28) p. xiii.

38   Crookes to Stokes, 23 September 1873.

39   Crookes to Stokes, 15 November 1873 *CUL Stokes Collection* C1078.

40   Crookes to Stokes, 25 September 1873 *CUL Stokes Collection* C1077.

41   Crookes to Stokes, 14 April 1874 *CUL Stokes Collection* C1082. Crookes used his own title in a report on his results that he sent to the *Philosophical Magazine:* 'On Attraction and Repulsion Accompanying Radiation' *Phil. Mag.* **48** (1874) 81–95.

42   Stokes to Crookes, 18 April 1876 *Memoir* II 372–3; Crookes to Stokes, 19 April 1876 *CUL Stokes Collection* C1091 ('I am scarcely yet satisfied that the residual gas is the cause.'); and Crookes to Stokes, 9 November 1876 *CUL Stokes Collection* C1111 ('I think I have absolute demonstration of the truth of the residual gas theory.')

43   Stokes to Crookes, 20 April 1876 *Memoir* II 373–4.

44   Stokes to Crookes, 20 January 1877, 19 July 1878, 13 June 1878 *Memoir* II 391–3, 398–9, 405–6, and Crookes to Stokes, 24 June 1878 *CUL Stokes Collection* C1129, when Crookes thought his experiments on viscosity and on the dissipation of heat in vacua would keep him busy until Christmas.

45   Crookes 1881 'On Heat Conduction in Highly Rarefied Air' *Proc. R. Soc.* **31** 239–43; Crookes 1881 'On the Viscosity of Gases at High Exhaustions' *Phil. Trans.* **172** 387–434; and Stokes 'Note on the Reduction of Mr. Crookes's Experiments on the Decrement of the Arc of Vibration of a Mica Plate Oscillating within a Bulb Containing More or Less Rarefied Gas' *MPP* V 100–16.

46   Crookes to Stokes, 1 October 1878, 2 October 1878, 6 October 1878 *CUL Stokes Collection* C1133, C1134, C1135.

47   Stokes to Crookes, 17 October 1878 *Memoir* II 410.

48   Crookes 'On the Illumination of Lines of Molecular Pressure' (note 30).

49   Crookes to Stokes, 18 November 1878 *CUL Stokes Collection* C1140.

50   Crookes 1878 'On the Illumination of Lines of Molecular Pressure, and the Trajectory of Molecules' *Proc. R. Soc.* **28** 110–11.

51   Stokes's referee's report (note 30): 'The most important [remark submitted to the author] relates to what I regard as a false analogy between a ray of light in the ordinary sense of the words and what the author, not I think very judiciously, calls a ray of molecular light.' A portion of Stokes's detailed comments to Crookes is printed as 'Extracts from Remarks of Detail on Mr Crookes's Paper, of Dec. 3, 1879 [should be Dec. 5, 1878], "On the Illumination of Lines of Molecular Pressure, and the Trajectory of Molecules"' *Memoir* II 421. The extract does not mention rays of molecular light.

52   Crookes to Stokes, 20 January 1879 *CUL Stokes Collection* C1144.

53   Stokes to Crookes, 25 January 1879 *Memoir* II 423.

54   Crookes 'On the Illumination of Lines of Molecular Pressure' (note 30), p. 164. In the abstract this passage read: 'Where the corpuscular theory of light holds good, and where light does not always move in a straight line.' Crookes 'On the Illumination of Lines of Molecular Pressure' (note 50) p. 111.

55   Stokes to Crookes, 21 October 1878 *Memoir* II 413.

56   *Ibid.* II 412.

57   Stokes to Crookes, 17 October 1878 *Memoir* II 411.

58   Crookes to Stokes, 28 October 1878 *CUL Stokes Collection* C1137, and Crookes 'On the Illumination of Lines of Molecular Pressure' (note 30) pp. 150–2. In addition, Crookes found that the film, but not the plate, would lose its phosphorescence after a period of time, probably, Crookes thought, because it heated faster than the thicker plate. Agreeing, Stokes regarded the result as 'a perfect demonstration that the phosphorescence is not due to radiation' (Stokes to Crookes, 29 October 1878 *Memoir* II 417, and *CUL Stokes Collection* C1138. The printed version of this letter is incomplete. Crookes incorporated much of Stokes's letter into his published paper.)

59   This is nicely done by David L Anderson in his *The Discovery of the Electron: the Development of the Atomic Concept of Electricity* (Princeton: Van Nostrand, 1964) pp. 21–47.

60   Crookes 'On the Illumination of Lines of Molecular Force' (note 30) p. 143n, and Crookes's footnote to H Eugen Goldstein 1879 'On Electric Discharges in Attenuated Gases, and on Certain Phenomena in Geissler's Tubes' *The Chemical News* **39** (30 May) p. 231n.

61   J Plücker 'On the Action of the Magnet upon the Electrical Discharge in Rarefied Gases' *Phil. Mag.* **16** (1858) 119–35, 408–18 **18** (1859) 1–7; J Plücker 1859 'Observations on the Electric Discharge' *Phil. Mag.* **18** 7–20; W Hittorf 'Ueber die Elektricitätsleitung der Gase' *Annalen der Physik und Chemie* **136** (1869), 1–31, 197–234, *Annales de Chimie et de Physique* **17** (1869), 487–96; E Goldstein 1875 'On Spectra of Gases' *Phil. Mag.* **49** 333–45; and E Goldstein 1877 'Preliminary Communication on Electric Discharges through Rarefied Gases' *Phil. Mag.* **4** 353–63. In a paper presented in 1878 but not published until 1880, Goldstein declared that cathode rays 'must be placed side by side with

the well-known forms of vibration which constitute light-waves and sound-waves'. ('On the Electric Discharge in Rarefied Gases' *Phil. Mag.* **10** (1880) 173.) In 1883, Heinrich Hertz attributed the ether-wave theory of cathode rays, which he himself accepted, to the later German Physicists, Goldstein and E Wiedemann, not the earlier ones, Plücker and Hittorf. (Hertz 'Experiments on the Cathode Discharge' in *Miscellaneous Papers* introduction by P Lenard, translated by D E Jones and G A Schott (London: Macmillan and Co., 1896) p. 246.

62 A Private Deschanel 1877 *Elementary Treatise on Natural Philosophy* 4th edn translated and edited with extensive additions by J D Everett (London: Blackie and Son) pp. 549–53, 765–6.

63 Stokes to Crookes, 20 April 1876 *Memoir* II 374. G Johnstone Stoney 'On Crookes's Radiometer' *Phil. Mag.* **1** (March 1876) 177–81, and 'On Crookes's Radiometer—Part II' *Phil. Mag.* **1** (April 1876) 305–13. On Crookes's critics, see DeKosky 'William Crookes and the Fourth State of Matter' (note 28) pp. 58–9.

64 Crookes 1876 'On Repulsion Resulting from Radiation. Influence of the Residual Gas. Preliminary Notice' *Proc. R. Soc.* **25** (15 June) 139–40, and 'Experimental Contributions to the Theory of the Radiometer' *Proc. R. Soc.* **25** (16 November) 308n.

65 Stokes to Crookes, 21 October 1878 *Memoir* II 412.

66 J J Thomson 1893 *Notes on Recent Researches in Electricity and Magnetism* (Oxford: Clarendon) p. 126.

67 C F Varley 1871 'Some Experiments on the Discharge of Electricity through Rarefied Media and the Atmosphere' *Proc. R. Soc.* **19** 239.

68 Stokes 'communicated' Varley's paper to the Royal Society (*ibid.* p. 236), and Crookes was influenced in the late 1860s by Varley's interest in spiritualism (Fournier D'Albe *Sir William Crookes* (note 28) p. 133).

69 Kelvin to Stokes, 1 February 1896 *CUL Stokes Collection* K307.

70 Stokes to Kelvin, 10 February 1896 *CUL Stokes Collection* S499.

71 *Ibid.*

72 *Ibid.*

73 Kelvin to Stokes, 12 February 1896 *CUL Stokes Collection* K309, *Thompson* II 957. In October and November of 1896, Kelvin read Hertz's 1883 paper on cathode rays in his own copy of Hertz's *Miscellaneous Papers* (note 61), published earlier that year. He heavily annotated it with critical comments: 'utterly invalid' (p. 232), 'Varley long ago, and we in July & Sep. 1896 have utterly disproved this' (p. 236), 'very bad' (p. 245), 'why not the Varley-Crookes torrent?' (p. 246), 'This is astonishing. It shows that in 1883 Hertz did not know electro statics; and helps us understand how he could fail to see what his observation of p. 251 11 13 . . . 16 proved' (p. 250). Kelvin dated each remark. The volume is in the Kelvin Library in the Department of Natural Philosophy, Glasgow University.

74 Stokes to Crookes, 24 October 1896 *Memoir* II 474.

75 Stokes to Crookes, 31 March 1897 *Memoir* II 476.

76 Stokes to Kelvin, 26 March 1896 *CUL Kelvin Collection* S510.

77   Stokes to Kelvin, 31 March 1896 *CUL Kelvin Collection* S512.

78   Stokes '[The Wilde Lecture] On the Nature of Röntgen Rays' *MPP* V 256–77. The earlier lectures were 'An Annual Address (Chiefly on the Subject of the Röntgen Rays)' *J. Trans. Victoria Inst.* **30** (15 July 1896) 13–27, and 'On the Nature of the Röntgen Rays' *MPP* V 254–5, delivered on 9 November 1896 to the Cambridge Philosophical Society.

79   B R· Wheaton 1983 *The Tiger and the Shark: Empirical Roots of Wave-Particle Dualism* (Cambridge: Cambridge University Press) pp. 15–48.

80   J J Thomson 1896 'The Röntgen Rays' *Nature* **54** (30 July) 304.

81   Stokes 'Wilde Lecture' *MPP* V 273–4.

82   Crookes 1898 'Address' *Br. Assoc. Rep.* 27.

83   Stokes to Crookes, 19 April 1900 *Memoir* II 478.

84   Crookes to Stokes, 21 April 1900 *Memoir* II 478.

85   Stokes to Crookes, 23 April 1900 *Memoir* II 478.

86   Crookes to Stokes, 27 April 1900 *Memoir* II 480.

87   Stokes to Crookes, 1 May 1900 *Memoir* II 480–1.

88   Stokes to Becquerel, 16 August 1899 *CUL Stokes Collection* B242; *Memoir* I 293.

89   Stokes to Becquerel, 4 September 1899 *CUL Stokes Collection* B243; *Memoir* I 298.

90   Becquerel to Stokes, 25 August 1899 *CUL Stokes Collection* B244; *Memoir* I 294–7.

91   Stokes to Crookes, 15 December 1900 *Memoir* II 482. Strutt's research in Cambridge which involved this aspect of Becquerel rays was reported in a paper received by the Royal Society on the same day Stokes wrote to Crookes: R J Strutt 1901 'On the Conductivity of Gases under the Becquerel Rays' *Phil. Trans.* A **196** 507–27.

92   Stokes to Crookes, 15 December 1900 *Memoir* II 482. 'Rays' and 'jets' were part of a new nomenclature suggested by Stokes to Crookes.

93   Crookes to Stokes, 16 December 1900 *Memoir* II 484. Earlier Crookes had regarded Röntgen rays as waves. (Crookes 'Address' (note 82) p. 24.)

94   Smithells to Larmor (note 23) p. 266.

95   T J Trenn 1976 'Rutherford on the Alpha-Beta-Gamma Classification of Radioactive Rays' *Isis* **67** 61–75.

9

Radioactivity:
Kelvin versus Rutherford

Few scientific careers have been so successful so quickly as that of Ernest Rutherford. Entering the mainstream of European physics in J J Thomson's Cavendish Laboratory in 1895, he won his Nobel prize only thirteen years later in 1908. Born and educated in New Zealand, Rutherford was at the Cavendish for three years before accepting a professorship at McGill University in Montreal. Though Rutherford had done important work on radioactivity in Cambridge—distinguishing between and naming the alpha- and beta-radiations—it was in Canada from 1901 to 1903 that he teamed with the Oxford-educated chemist, Frederick Soddy, to work out the theory of the transmutation of the elements. According to this theory, an atom, upon emitting an alpha- or beta-ray, 'disintegrated' into an atom of another element. Published in 1904, Rutherford's book, *Radio-Activity* (second edition 1905), reflected his leading position in the field. In 1905 he was invited to deliver a series of Silliman lectures at Yale University, which became the basis for another book, *Radioactive Transformations*, published in 1906. Returning to England in 1907, he headed the physics laboratory at Manchester University where he established in 1908 that alpha-particles were positively charged helium atoms, a result he had strongly suspected for five years. In connection with scattering experiments with alpha-particles, he published his theory of the nuclear atom in 1911. After the First World War, Rutherford replaced Thomson at the Cavendish, maintaining it as a centre of experimental physics during the 1920s and early 1930s. This chapter, however, concerns only Rutherford's early career, for that, obviously, was the period of Kelvin's response to Rutherford's new ideas. Indeed, the elder statesman of British physics was the rising young star's chief critic.

There already exists one explanation of Kelvin's opposition to Rutherford's ideas—a combination of advanced years and meagre knowledge. The explanation possibly originated with Rutherford himself, as sometimes happens when a victor comments on the history of his area of science. A S Eve's 1939 biography of Rutherford unmistakeably conveys this interpretation of the Kelvin–Rutherford dispute. For example, Eve quotes Rutherford writing to his fiancée about Kelvin in 1896: 'I hope to see the old fellow before his days are done, as he is very old now, although he is still very active and works away as hard as ever it seems. . . .'[1] In Rutherford's entertaining reminiscence of his 1904 lecture on radioactivity at the Royal Institution to an audience containing Kelvin, he refers to Kelvin as the 'old bird' and the 'old boy'.

> I came into the room, which was half dark, and presently spotted Lord Kelvin in the audience and realized that I was in for trouble at the last part of the speech dealing with the age of the earth, where my views conflicted with his. To my relief, Kelvin fell fast asleep, but as I came to the important point, I saw the old bird sit up, open an eye and cock a baleful glance at me! Then a sudden inspiration came, and I said Lord Kelvin had limited the age of the earth, *provided no new source of heat was discovered.* That prophetic utterance refers to what we are now considering tonight, radium! Behold! the old boy beamed upon me.[2]

Shortly after the lecture, Kelvin and Rutherford were among Lord Rayleigh's guests at his home in Essex. Rutherford wrote to his wife in Canada:

> Lord Kelvin has talked radium most of the day, and I admire his confidence in talking about a subject of which he has taken the trouble to learn so little. I showed him and the ladies some experiments this evening, and he was tremendously delighted and has gone to bed happy with a few small phosphorescent things I gave him.[3]

Variations on the theme of the old-and-ill-informed Kelvin have been regularly repeated—recently in the article on Rutherford in the *Dictionary of Scientific Biography*: 'Aside from the elderly and unalterable Lord Kelvin and the constantly contentious Henry Armstrong, the transformation theory encountered little opposition'[4] and more recently in a history of atomic physics: 'Kelvin's objections could be ascribed to the vagaries of extreme old age.'[5] Perhaps tacitly accepting the established view, authors of studies on Kelvin tend to avoid his work on radioactivity.[6] The view's validity would seem only to be underscored by the close proximity of the 83-year-old's death in 1907 to the 37-year-old's winning of a Nobel prize the following year.

Consequently, Kelvin's ideas about radioactivity and atomic structure may appear a stern test for this book's historiographical point of view. After all, was not Kelvin simply *wrong* in opposing the widely supported theories of Rutherford and his teacher J J Thomson—theories which have so successfully led to so much of modern physical theory? How can Kelvin be seen, even with reference to his own context, to be 'right'? Was not Kelvin *so* wrong that an age-induced rigidity or confusion of thought is the obvious explanation? Maybe. However, even if such a judgement were correct, there would remain the historical task of identifying Kelvin's ostensible reasons for opposing Rutherford, certainly an appropriate task for a study devoted to Kelvin. More than that, though, the chapter contends that Kelvin's essential arguments constituted a not indefensible position, one rooted in a Victorian natural philosopher's comprehensive consideration of energy relations, Boscovichian forces, theological realities, atomic structure, and the ages of the earth and sun. At the least, the chapter helps point out some of the uncertainties in early-twentieth-century physics. In doing so, it by no means detracts from the accomplishments of Thomson and Rutherford, but perhaps puts those accomplishments in better perspective.

The Thomson–Rutherford research programme, 1897–1907

Though by no means the only ones working with electrons, radio-activity and atomic structure, J J Thomson and Rutherford were the most prominent. Thomson's 1897 discovery of the electron, or 'corpuscle' as he called it, and Rutherford's research on radioactivity contributed fundamentally to Thomson's so-called 'plum pudding' model of the atom, proposed in his 1903 Silliman lectures at Yale. That model, in turn, provided the theoretical framework for guiding and understanding subsequent research. Thomson's Silliman lectures were published as *Electricity and Matter* in 1904, and three years later he published *The Corpuscular Theory of Matter*. Those two books combined with Rutherford's to define the field. In taking exception to the Thomson–Rutherford viewpoint, for example, Kelvin relied on their books (especially *Electricity and Matter* and *Radio-Activity*) for pertinent information.[7]

At the outset, we should emphasise that Kelvin did influence the mainstream research of Thomson and Rutherford from which he himself diverged. In the first place, Thomson and Rutherford regarded their atomic models in the tradition of Kelvin's model-building approach to physical theory. The immediate question was not whether the models were true, but whether they were fruitful in aiding thought and guiding research. Such models, Thomson wrote,

> may be useful even though the models are crude, for if we picture to ourselves how the model atom must be behaving in some particular physical or chemical process, we not only gain a very vivid conception of the process, but also often suggestions that the process under consideration must be connected with other processes, and thus further investigations are promoted by this method.[8]

Second, the idea of positive electricity diffused uniformly through the atom, an essential part of Thomson's plum-pudding atomic model, was first suggested by Kelvin. Wrote Thomson: 'Let us picture to ourselves the aggregate as, like the AEpinus atom of Lord Kelvin, consisting of a sphere of uniform positive electrification. . . .'[9] Though embracing at least one of Kelvin's specific ideas, Thomson and Rutherford essentially disagreed with him. But it was disagreement within a common methodological framework which duly noted the provisional quality of physical models. Thomson pointed out, for example, that in the absence of any firm knowledge of the actual distribution of positive electricity within the atom, he assumed a uniform distribution because it was the simplest mathematically.

The summer of 1903, just after Thomson's Silliman lectures and just before Kelvin's first paper on radioactivity, is a convenient time for summing up progress made within the Thomson Rutherford programme. Thomson's idea that the electron or corpuscle was much smaller than an atom had been confirmed in various ways. A third radiation, gamma-rays, had been added in 1900 to the two Rutherford had earlier identified. Beta-rays had been deflected in electric and magnetic fields in 1900, showing that they were electrons moving at high speeds. Rutherford had finally succeeded in similarly deflecting alpha-rays in 1902, revealing them to be positively charged particles with a mass of the order of that of hydrogen or helium atoms. Still undeflected, gamma-rays seemed akin to x-rays. The Curies had identified radioactive elements in addition to Becquerel's uranium, and Marie Curie had managed to isolate radium in 1902. In the spring of 1903 Pierre Curie published his startling finding that radium produced enormous quantities of heat. By the autumn of 1903, Rutherford and H T Barnes had shown that the kinetic energy of alpha-particles produced most of the heat. Also published in the spring of 1903, Rutherford and Soddy's disintegration theory sought to explain a mass of experimental results concerning various radioactive substances and their radiations. As the amount of one element decreased, they concluded, the amount of another increased. One was transmuting into the other as it emitted its radiation. Thomson's plum-pudding atom offered a provisional picture at the atomic level. Within a sphere of positive electricity, thousands of corpuscles moved rapidly in a series of concentric orbits. The corpuscles constituted virtually all of the mass of the atom. Emission of alpha-particles indicated, Thomson wrote, 'either that radium is a compound containing lighter elements or else that the atom of radium is disintegrating into such elements'.[10] Developing the disintegration theory, Thomson pointed out that his atom would be stable for a long period of time because the electromagnetic radiation from any one orbiting corpuscle would be mostly absorbed by others within the atom. However, the gradual loss of radiation would eventually produce an unstable situation that might precipitate radioactive disintegration.

By 1907 Thomson's model had been helpful. Most significant, it had guided Thomson's interpretation of beta-scattering experiments. Assuming that the overall scattering resulted from the beta-particle's multiple interactions with the atom's corpuscles, where the charge and mass within the atom were concentrated, Thomson inferred that the number of corpuscles within the atom must be of the same order as the atom's atomic number, vastly fewer than previously supposed. That, of course, meant that the positive electricity must carry a proportionately greater share of the atom's mass. It also meant there was a potentially serious problem. Some, though not Thomson, feared that the many fewer electrons would be able to absorb much less of the electromagnetic radiation produced by their fellows, thus destroying the plum-pudding atom's long-term stability. Discounting that potential difficulty, Thomson instead emphasised that the requirement of fewer electrons strengthened the model 'since it makes the number of possible atoms much more nearly equal to the number of the chemical elements'.[11]

Looking ahead to the contrast between this programme and Kelvin's, we need to examine three questions. What was the source of the alpha- and beta-particles' great kinetic energy? What exactly happened within the atom at the moment of disintegration? How did the fact that radioactive substances emitted large amounts of heat affect ideas about the ages of the earth and sun? Interrelated, all three questions involved energy considerations in one way or another.

Though opposed by proponents of various 'external' theories, Thomson and Rutherford favoured an 'internal' theory of the source of the particles' kinetic energy. The energy, that is, came entirely from within the atom and was not absorbed from outside. Supporters of 'external' theories thought this an inconceivable amount of energy to reside within an atom. However, Thomson and Rutherford pointed out, the rate of radioactivity remained undiminished in various experiments where radioactive substances were isolated from possible external sources of energy.

Thomson and Rutherford could not be extremely precise about what happened within the atom during disintegration. In 1904 Thomson suggested that the eventual instability within the atom would lead to a new configuration of corpuscles. Like a spinning top slowing down, the atom would possess a critical value of kinetic energy at which it would crash. 'When it reaches this value the crash comes, the original configuration is broken up, there is a great decrease in the potential energy of the system accompanied by an equal increase in the kinetic energy of the corpuscles. The increase in the velocity of the corpuscles may cause the disruption of the atom into two or more systems. . . .'[12] One system would be an atom with about the mass of the original atom. An alpha-particle would be a system of a few thousand corpuscles within their positive matrix, and a beta-particle would be a system of a single corpuscle. Rutherford also emphasised the kinetic energy of the atom's electrons prior to ejection from the atom. He wrote in 1904, for example:

> It seems very improbable that α and β particles suddenly acquire their enormous velocity of projection by the action of forces existing inside or outside the atom. For example, the α particle would have to travel from rest between two points differing in potential by 5.2 million volts in order to acquire the kinetic energy with which it escapes. Thus it seems probable that these particles are not set suddenly in motion, but that they escape from an atomic system in which they were already in rapid oscillatory or orbital motion. On this view the energy is not communicated to the projected particles, but exists beforehand in the atoms from which they escape.[13]

The fact of radium's heat was quickly made relevant to cosmological issues. As we saw in Chapter 1, Kelvin had limited the ages of the earth and sun through the coinciding conclusions of three sets of calculations, involving the gravitational source of the sun's heat, the tidal retardation of the earth's spin and the continual dissipation of the earth's heat. If radium had contributed to the heat of the sun and earth, that would alter Kelvin's estimates, greatly increasing the ages of the sun and earth. This point was argued in 1903 and 1904, most prominently by Rutherford himself, who concluded his 1904 Royal Institution lecture on precisely this subject:

> Thus the earth may have been at a temperature capable of supporting animal and vegetable life for a much longer time than estimated by Lord Kelvin from thermal data. Similar considerations apply to the question of the sun's heat; for the presence of radium in the sun, to the

extent of about four parts in one million by weight, would of itself account for the present rate of emission of heat. The discovery of the radio-active elements, which in their disintegration liberate enormous amounts of energy, thus increases the possible limit of the duration of life on this planet, and allows the time claimed by the geologist and biologist for the process of evolution.[14]

Finally, we should realise that some familiar parts of modern science were *not* so important from 1897 to 1907. Radiometric dating techniques, which now provide accurate ages independently of considerations of the cooling of the earth, barely existed before 1907. Rutherford and others did make such measurements, but the estimates of half-lives and the details of radioactive decay series were too uncertain for the measurements to be influential.[15] That is, radioactivity's principal objection to a young earth rested on rate of cooling not on radiometric dating. Einstein published his prediction of the relativistic conversion of mass to energy at the end of the period, but it played no part in British discussions of radioactivity. Only much later would relativity be seen to provide an explanation of the heat of radium and, in conjunction with the fusion of hydrogen into helium, that of the sun. Kelvin and Rutherford were therefore equally committed to the nineteenth-century version of the conservation of energy.

Kelvin's research programme, 1897–1907

The commencement of Kelvin's modern research programme on atomic structure coincided with J J Thomson's discovery of the electron, but evidently did not depend on it. Nearly simultaneously with Thomson's announcement of corpuscles in the spring of 1897, Kelvin proposed his idea of 'electrions' in order to explain ordinary electrical phenomena like contact electricity, electrolysis, and the difference between conductors and insulators. Accepting evidence for the existence of a fundamental unit of electric charge, Kelvin presented an atomistic version of the one-fluid electrical theory of the eighteenth-century natural philosopher, Franz Aepinus. Combining the established term *ion* with the term suggested in 1892 for the unit charge, *electron*, an electrion was an atom of resinous, or negative, electricity. Matter without its proper share of electrions would behave as vitreous, or positive, electricity. A Boscovichian-type force linked resinous and vitreous electricity. Kelvin's 1897 paper began a series which appeared over the next decade. In 1901 he completed the most well known paper in the series, 'Aepinus Atomized'. Still focusing on ordinary electrical phenomena, Kelvin now equated his electrions with Thomson's corpuscles. In turn, Thomson drew his idea for the plum-pudding atom from 'Aepinus Atomized' and later included some of Kelvin's conclusions in *The Corpuscular Theory of Matter*.[16] The prominent Dutch physicist, H A Lorentz, after reading 'Aepinus Atomized', wrote to Kelvin that 'we have now at least a possibility of understanding some phenomena that were till now wholly mysterious'.[17] Kelvin's 1901 atom was a static arrangement of its constituent parts. Circumstances could destroy the atom's equilibrium, however, with the resultant motions of electrions causing observable electrical phenomena. Later, taking note of experimental results of Rutherford and others, Kelvin adapted his atom of static arrangement and potential energy to the facts of radioactivity. In 1904 he developed an atomic model in which the same atomic process yielded both an alpha- and a beta-particle. Concentrating on the emission of beta-particles in 1905, he constructed a different model which incorporated a more elaborate Boscovichian force curve than had his earlier models. This model he continued to refine through his final year. Hence, though responding to certain developments associated with the Thomson–Rutherford programme, Kelvin's mainly followed out his own theoretical insights. Thus retaining its essential independence, it also enjoyed some recognised successes.

Among other things, as we can see, Kelvin's atomic models were part of the last phase of his long consideration of Boscovich's theory. As discussed in Chapter 7, Kelvin had rejected the Boscovichian notion of infinitesimal atoms by the early 1860s, as he earlier had the Boscovichian idea of action at a distance. In 1884 he contrasted the Boscovichian atom unfavourably with the vortex atom.[18] By the early 1890s, however, he was readily employing Boscovichian force curves as suggestive ways of analysing the structure and properties of crystals. 'Without accepting Boscovich's fundamental doctrine . . . , we can learn something towards an understanding of the real molecular structure of matter, and of some of its thermodynamic properties, by consideration of the static and kinetic problems which it suggests.'[19] It was a helpful, though 'infinitely improbable theory', as he stated in 1893.[20] With the continued failure of the vortex-atom theory, however, Kelvin's view of Boscovich had changed by the early years of the twentieth century. He noted in 1900, for example, that the vortex-atom's potential was 'zero'.[21] Evidently, failure of pure matter in motion caused, or at least contributed to, Kelvin's adoption of the only alternative for microscopic reality, action at a distance, with the proper form of the force curve to be determined, in the style of Boscovich, by experience. One postulates the existence of whatever force curve is necessary to fit with experimental results. Natural phenomena, Kelvin wrote in the *Baltimore Lectures* of 1904, were 'to be explained by a proper law of force according to the Boscovichian doctrine which we all now accept (many of us without knowing that we do so) as the fundamental hypothesis of physics and chemistry'.[22]

Scheduled to give an address at the Royal Institution on 21 May 1897 on the controversial topic of contact electricity, Kelvin asked himself on 13 May, 'What can be the difference betw. Zn & Cu atoms?'[23] At issue was the cause of voltaic electricity, the subject of continuing disagreement through the century. On the basis of his and Joule's experiments in the 1850s, Kelvin had argued that, because of the chemical affinity of the two for one another, contact between zinc and copper alone sufficed to produce voltaic electricity. Others had argued that chemical action, involving the surrounding oxygen according to Oliver Lodge, was the cause. Rambling through his Royal Institution lecture, Kelvin never got around to his atomic theory of contact electricity and so sent a short paper the following day to *Nature*. Figure 9.1 portrays the theory. 'Adopting the essentials of Aepinus' theory, and dealing with it according to the doctrine of Father Boscovich', Kelvin wrote, 'each atom of ponderable matter is an electron of vitreous electricity [C and C'] . . . with a neutralizing electrion of resinous electricity [the *es*] close to it.'[24] The electrions repel one another according to Coulomb's inverse-square law. At distances greater than r and r', respectively, C and C' repel the electrions according to Coulomb's law. However, for distances less than r and r', the force becomes zero or perhaps strongly repulsive. In insulators, the closest approach of two atoms to one another will be greater than twice their radius, in conductors less than twice. In figure 9.1, two different atoms (say of zinc and copper) are in contact, with the two electrions in positions of stable equilibrium. If the atoms are separated from one another, either by electrolytic forces as in figure 9.1 or mechanically as when two metals in contact are pulled apart, the electrions will stay with the smaller atom, C'. 'Two pieces of metal, M, M' . . . will behave in respect to contact-electricity just as two pieces of metal behave in a perfect vacuum. For example, if $r > r'$, M will behave to M' as zinc behaves to copper'.[25]

In 'Aepinus Atomized' of 1901, Kelvin simply expanded the 1897 atoms, C and C', into the atoms, A and A', containing uniformly distributed vitreous electricity (figures 9.2 and 9.3). Here, Kelvin employed only Coulomb's law. Hence, the attraction of the vitreous A for an electrion outside the sphere defined by r would be inversely proportional to the square of the distance between C and the electrion. Within the sphere, the attraction would accordingly decrease with the distance, becoming zero at the atom's centre. The electrion therefore finds stable equilibrium only at the atom's centre as shown in

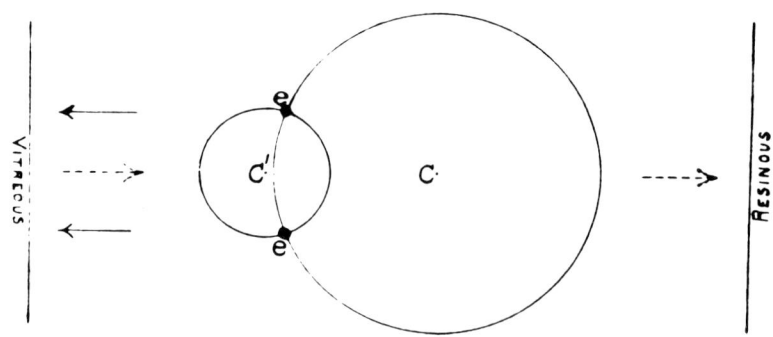

Figure 9.1 From Kelvin 1897 'Contact Electricity and Electrolysis According to Father Boscovich' *Nature* **56** (27 May) 84.

figure 9.2 where the attraction and repulsion between A' and E', respectively, and E offset one another. However, when the centre of A penetrates the sphere of A', as in figure 9.3, the network of forces shifts the equilibrium points for E and E' away from the centres of their respective atoms. Furthermore, if A and A' were first to be made concentric with one another and then to be separated, the two electrions would end up in the smaller atom, in stable positions equidistant from C. A and A' would thus be oppositely charged. If, instead of these 'mono-electrionic' atoms, the process involved two 'poly-electrionic' atoms, the smaller atom would end up with the majority of the electrions. 'This is a very remarkable conclusion, pointing to what is probably the true explanation of the first known of the electric properties of matter; attractions and repulsions produced by rubbed amber.'[26] To illustrate further the tendency of smaller atoms to capture electrions from larger ones, Kelvin considered an interaction between a large atom with one electrion and a small atom with none. Under the right circumstances, the large atom's electrion would reach unstable equilibrium and then 'jump out' of the atom 'like a cork jumping out of a bottle', eventually coming to rest within the small atom after several oscillations around the final resting place.[27]

Kelvin went on in 'Aepinus Atomized' to discuss several aspects of polyelectrionic atoms. What arrangements of electrions, for example, would provide stable equilibrium? 'For any number of electrions there may be equilibrium with all the electrions on one spherical surface concentric with the atom.'[28] Moreover, 'from any case of any number of electrions all on one spherical surface, we may pass to another configuration with one more electrion placed at the centre and the proper proportionate increase in the electric strength of the atom'.[29] The more electrions such an atom contained, Kelvin argued, the easier it would be to dislodge them from the atom. Thus, raising the temperature of a substance composed of such atoms 'sets the electrions to performing wildly irregular vibrations and rotations, so that some of them will occasionally be shot out of their atoms'.[30] This was the 'matter-of-fact explanation' of certain experiments indicating that electrical conductivity depended on temperature.

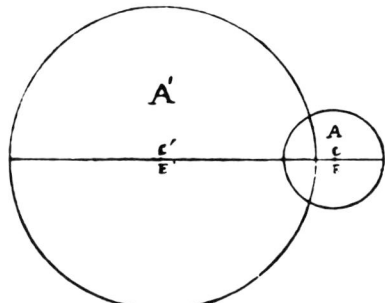

Figure 9.2 From Kelvin 1902 'Aepinus Atomized' *Phil. Mag.* **3** (March) 261.

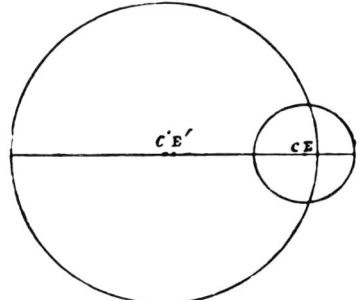

Figure 9.3 From Kelvin 1902 'Aepinus Atomized' *Phil. Mag.* **3** (March) 261.

Kelvin's 1901 atomic model underlay his tentative account of radio-activity in 1903. Unable to attend the British Association meeting that year, he sent a paper proposing 'my atomic resuscitation of the old doctrine of Aepinus' as an explanation. 'The ordinary thermal motions within any solid, or liquid, or gas, must cause occasional shootings out of the electrions from the substance. . . .'[31] Hence, all substances must be radioactive to some degree, and radium was so intensely radioactive probably 'because it is exceedingly poly-electrionic'.[32] Doubting that radium atoms could store enough energy to produce the intense heat reported by Curie, Kelvin regarded the radium atom as a system somehow capable of more or less immediately transforming energy of heretofore undetected ethereal waves into the emitted heat. As a test for this view, he proposed an experiment to compare 'the thermal emission from radium wholly surrounded with thick lead with that found with the surroundings hitherto used'.[33]

By the next British Association meeting in August 1904, Kelvin had worked out new atomic models for radioactivity. By then he had been reading Rutherford's *Radio-Activity* which reported that a lead shield did not affect radium's thermal emission.[34] Moreover, as he later wrote to Henry Armstrong, he had been unable to imagine a mechanism by which the radium atom, analogous to the way quinine degraded ultra-violet light into visible light, could immediately transform external energy into its emitted energy. 'At present I can see nothing else than that the energy given out is taken from a previously existing store of potential energy of repulsive force between separable constituents of radium.'[35] Thus, he now adapted his concepts from 'Aepinus Atomized' to the atomic storage of enormous amounts of energy. Figures 9.4 and 9.5 are models for producing alpha-rays and beta-rays, respectively.[36] In figure 9.4, for example, the vitreous atoms have four times the charge of the resinous electrion. Assuming Coulomb's inverse-square law, simple calculations show that each of these arrangements is in equilibrium. The equilibrium is unstable, however, and each system verges on explosive separation of its parts. When that occurs, the electrion in figure 9.4 will settle into one of the two atoms, which violently repel one another becoming positively charged alpha-particles. Likewise, in figure 9.5, one of the electrions remains within the atom while the other is shot out as a negatively charged beta-particle.

These models had evident drawbacks. Since neither system in figure 9.4 or 9.5 was electrically neutral, attaining electrical neutrality required addition of other, oppositely charged atoms. Kelvin had to introduce a cluster of 14 negatively charged atoms as a 'neutralizing preservative guard' for the polonium molecule in figure 9.4, for example.[37] Second, because the energy of the explosion was inversely proportional to the sizes of the atoms in figures 9.4 and 9.5, achieving sufficient energy would probably demand atoms much smaller than atoms were known to be. Third, it was already being determined, as Rutherford recorded in *Radio-Activity*, that alpha-particles were only about as massive as hydrogen or helium atoms, vastly less than the mass of the polonium atoms in figure 9.4.

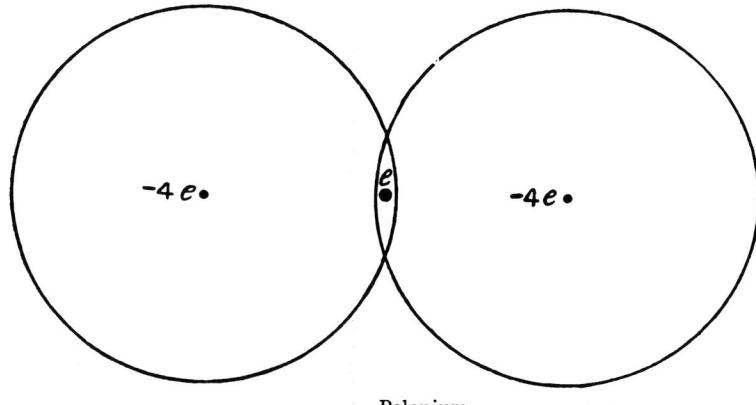

Polonium.

Figure 9.4 From Kelvin 'Plan of a Combination of Atoms Having the Properties of Polonium or Radium' *MPP* VI 217.

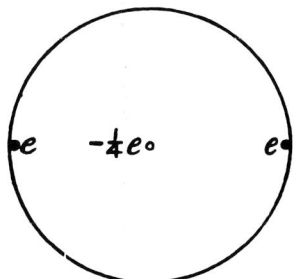

For β rays of Radium.

Figure 9.5 From Kelvin 'Plan of a Combination of Atoms Having the Properties of Polonium or Radium' *MPP* VI 217.

Whether contemplating these difficulties or others, Kelvin evidently awoke on 23 October 1905 with a superior model, at least for production of beta-particles. Illustrated in figure 9.6 from Kelvin's research notebook and in figure 9.7 from his published account, the model reintroduced a Boscovichian force curve.[38] Though Coulomb's inverse-square law applied at distances greater than the radius of an atom, that 'leaves us absolutely free to assume any law of force whatever that suits our purpose, when the electrion is within the atom'.[39] As drawn on the right-hand side of Kelvin's notebook entry for 23 October, the force curve represents a repulsive force when it is above the horizontal axis, attractive below. As Kelvin indicated in his notebook, the derivative of the curve for the electrion's potential energy yields the force curve. The potential curve is drawn at the top of the notebook entry and reproduced in figure 9.7. Thus, in figure 9.7, an electrion at the centre of the atom (i.e. at C) would experience no force, would possess a large amount of potential energy, would be attracted back towards C if it tended to move away and, therefore, would be in a position of stable equilibrium, where it could remain for a long time. Point M is another position of zero force, but a position of unstable equilibrium. An electrion there would tend to be either attracted back towards C or repelled with great force away from C. Like C, point N is a position of zero force and stable equilibrium, but an electrion there would possess slightly negative potential energy. The potential energy of an electrion at a particular point measured the work, positive or negative, required to move an electrion from an infinite distance to that point. Hence, in this atom, vitreous and resinous electricity were arranged in alternating concentric spheres, the resinous electricity consisting of electrions. Though the force was different, the stable geometrical configurations of electrions within the atom were similar to those in 'Aepinus Atomized'—a central electrion with others symmetrically arranged equidistant from it. Because of the structure of the force curve within the atom, electrions in a 'spherical surface' at N could leave or enter the atom relatively easily, with low energies involved, thus causing ordinary electrical phenomena. Because of the enormous repulsive force between M and N, however, it would be very difficult for an electrion to penetrate to the atom's centre. Once there, if repelled, it would be ejected with great energy as a beta-particle.

In his final paper on the subject published only a few months before his death, Kelvin refined these ideas and linked them to his ideas about the ether. Pursuing the consequences of his notebook entry for 24 October 1905 (figure 9.6), Kelvin postulated an atom possessing several concentric regions of alternating attractive and repulsive forces and, therefore, several spherical surfaces with electrions in stable equilibrium. Though electrions could move from place to place within an atom, they would eventually settle into an oscillatory motion at a surface of stable equilibrium. Because of the electrions' hold on the

Figure 9.6 (opposite) From Kelvin's research notebook *CUL Kelvin Collection* NB171. This is a page from one of Kelvin's 'green books' which he kept with him at all times. It records jottings from one week in October 1905, going from Kelvin's thoughts while lying in bed on a Monday morning to the posting of a completed paper the next Sunday to the *Philosophical Magazine*, where it appeared in the December issue. Figure 9.7 is from that paper. Figure 9.6 shows not only a potential-energy curve and the corresponding force curve, but also indicates Kelvin's contemplation of the more intricate potential curve within the atom which he did not publish until 1907. (By permission of the Syndics of Cambridge University Library.)

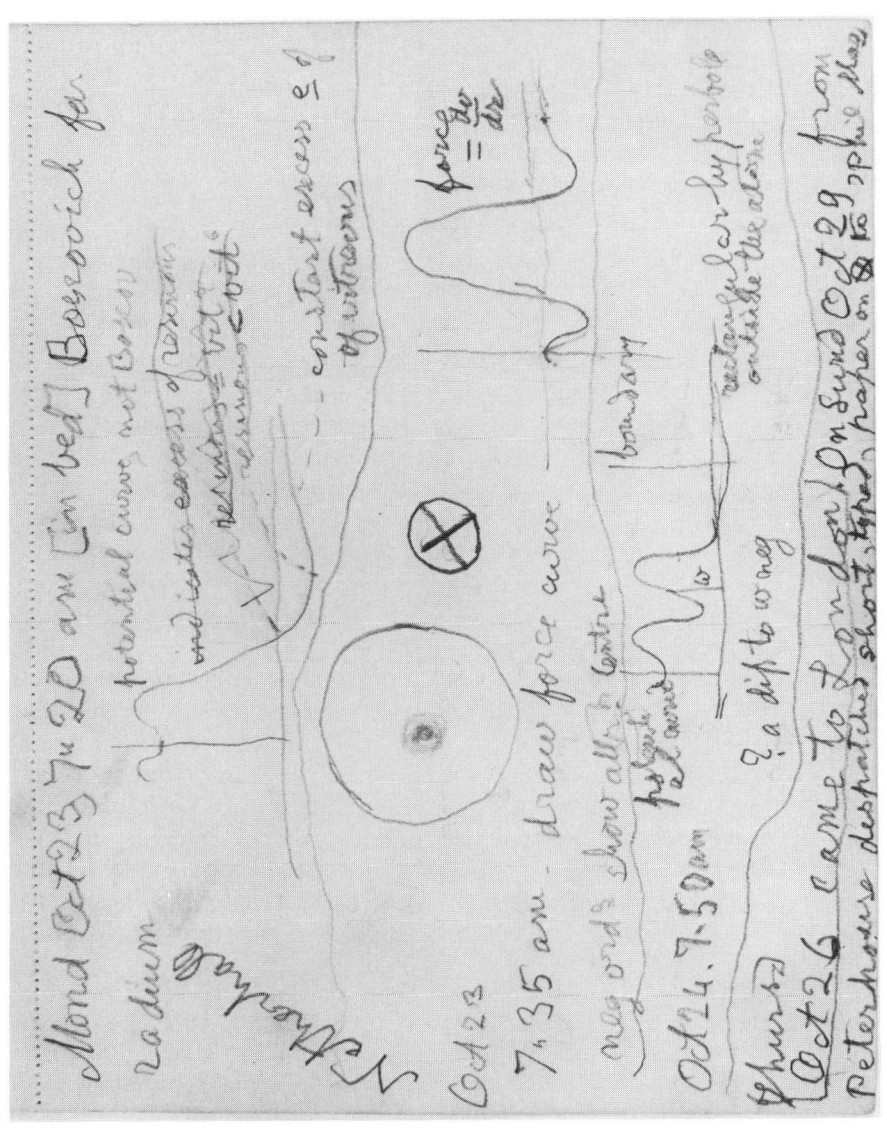

ether, their oscillations would produce ethereal waves. Hence, given many electrons and several surfaces of stable equilibrium per atom, one could now try to construct Boscovichian force curves that would tie interatomic electrionic vibrations to the growing number of empirical laws of bright-line spectra for the various chemical elements. In this paper, Kelvin still regarded alpha-particles as recoiling radium atoms. Concluding in support of an external theory of radium's energy, he thought it 'not absolutely impossible that Radium may be, as it were, an exceedingly black body, relatively to waves of ether so short that lead and other solid and liquid substances are transparent for them'.[40]

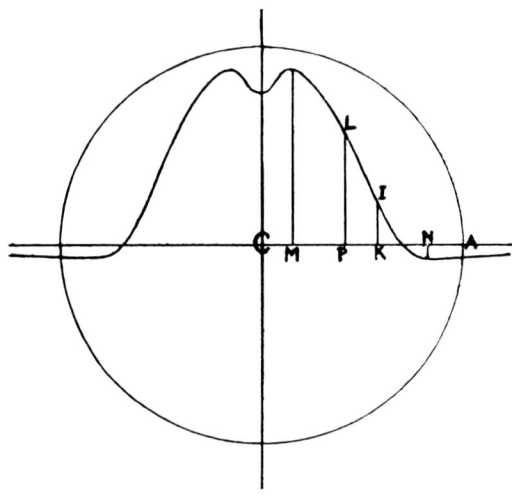

Work-curve.

On the right side of the diagram, slope up to right implies attraction;
slope down implies repulsion.

Figure 9.7 From Kelvin 'Plan of an Atom to be Capable of Storing an Electrion with Enormous Energy for Radio-Activity' *MPP* VI 228.

Once again, we should not overlook the common ground between these competing programmes. Kelvin would doubtlessly have agreed, for example, with the main thrust of Thomson's 1907 assessment: 'The corpuscular theory of matter with its assumptions of electrical charges and forces between them is not nearly so fundamental as the vortex theory of matter, in which all that is postulated is an incompressible, frictionless liquid possessing inertia and capable of transmitting pressure.'[41] The difference was that where Thomson merely saw insurmountable mathematical complexity in the vortex-atom theory, Kelvin by this time saw physical implausibility. Forced for different reasons to the second-best theory, necessitating action-at-a-distance forces, neither was absolutely bound to Coulomb's inverse-square force, which was, after all, only a particular form of a Boscovichian force curve. Kelvin readily used non-inverse-square forces. Thomson suggested a Boscovichian force curve as a way of defining the sizes of corpuscles' orbits within the atom. He noted that direct experimental verification of the Coulombic force curve existed only for distances and charges much larger than those within the atom.[42] In addition, as did Thomson and Rutherford, Kelvin usually noted the tentative quality of the specific features of his models. In 1903 he remarked that 'my suggestion respecting radium may be regarded as utterly unacceptable',[43] and in 1904 he pointed out that 'there are many other plans, some no doubt very much simpler than the combination of these two now suggested, for a combination of atoms to give the properties of Radium'.[44] Not all was tentative, however, and the next section focuses on specific points of conflict between Kelvin and Rutherford.

Kelvin versus Rutherford

The Kelvin–Rutherford dispute was a clash between two clusters of ideas, each cluster having a central concept to which its respective adherent was deeply committed. Rutherford's central concept was the disintegration theory of radioactivity; Kelvin's was the young-age theory of the earth.

Let us summarise Rutherford's position. By 1903 he and Soddy had published their striking theory of atomic disintegration, which became the theoretical centrepiece of Rutherford's views of radioactivity. He prefaced his 1904 volume: 'The interpretation of the results has, to a large extent, been based on the disintegration theory.'[45] The great energy released in radioactivity came from within the atom, none being absorbed from outside. The disintegration rate was constant. The internal energy existed primarily as the kinetic energy of the atom's parts. Once one atom disintegrated into another through emission of an alpha- or beta-particle, the new atom did not subsequently combine with a high-energy electron or helium atom to reverse the process. It was a dissipation of matter. Hence, Rutherford's *disintegration* theory entailed an *irreversible* release at a *constant rate* of the atom's *internal, kinetic* energy which *heated the earth*, thus increasing its calculated lifetime immensely.

Kelvin, according to J J Thomson's testimony, regarded his theory of the earth's age as his most important scientific contribution.[46] One can easily imagine its significance to Kelvin. Its publication had confronted prevailing mid-Victorian geological thought and had led Kelvin into vigorous debate with the outspoken biologist T H Huxley. It was part of that epic Victorian struggle between physics and biology, theism and naturalism. Though the overall war may not have been going so well, Kelvin had clearly won the specific battle about the earth's age, thus arming himself with solid argument in his continuing opposition to the intellectual fashions of the day. We can therefore readily appreciate Kelvin's deep conceptual–emotional commitment to the theory of a young earth. It was at least as crucial to Kelvin as the disintegration theory was to Rutherford, which promised, if generally accepted, to bring him high eminence. Moreover, we must remember that Kelvin had been working out his own atomic theory, largely independent of the Thomson–Rutherford programme. Given this combination of factors—unique to Kelvin among scientists of the century's first decade—it is scarcely surprising that Kelvin would instinctively oppose and then actively seek flaws in Rutherford's theory. In fact, he doubted not only the reality of disintegration, but also the many ideas connected with it.

Kelvin's early rejection of disintegration was unequivocal. 'The hypothesis of *evolution* of an atom seems to me utterly wild and improbable',[47] he wrote to James Dewar in August 1903. Even after considering the evidence over the next few years, however, he still doubted that radioactivity involved an *atomic* change. In August 1906 Kelvin wrote a letter to *The Times* which initiated an extended controversy in the pages of *The Times* and *Nature*. Objecting to the claim made at the recent British Association meeting that 'the production of helium from radium has established the fact of the gradual evolution of one element into others', Kelvin wrote that the finding

> suggests nothing more towards any modification of the atomic doctrine proposed some 2,500 years ago by Democritus and universally adopted by chemists and other philosophers in the 19th century, than does [William] Ramsay's original discovery of helium as an emanation from the mineral clevite. The obvious conclusion from the two discoveries is that clevite and radium both contain helium.[48]

In response to the answer that clevite and radium presented quite different circumstances, Kelvin replied that he agreed with the second edition of Rutherford's *Radio-Activity* 'that radium (atomic weight 225) may be a compound of four [*sic*, should be five] atoms of helium (atomic weight 4×5) and one atom of lead (atomic weight 205)'.[49]

Rutherford and his supporters wrote their own letters, Rutherford quoting extensively from *Radio-Activity*. Though radium contained helium, they said, the combination was definitely not an ordinary chemical compound. Chemically, radium behaved as an element, and like other elements it had a characteristic spectrum. Moreover, the enormous energy of alpha-particles meant that the process going on was not an ordinary chemical change.[50] Nevertheless, as Kelvin told Soddy, there remained the question of 'the propriety of using the terms atom and element in connexion with substances which had been shown by experiment to undergo change'.[51] The same point had been made two years earlier by the physicist H A Wilson in a review of Rutherford's and Soddy's books: 'If the radium atom consists of several parts which are separated during the process of disintegration, then the radium atom is really a molecule, and its disintegration is, strictly speaking, a molecular process.'[52] From Kelvin's perspective though radium may not have been an *ordinary* chemical compound neither was it an ordinary chemical *atom*.

In two letters to J J Thomson in 1906, Kelvin detailed his objections to the kinetic atom. Citing passages in the second edition of Rutherford's *Radio-Activity* as indication that the speed of beta-particles ranged up to values near and perhaps greater than the speed of light, Kelvin wrote:

> It does not seem to me improbable that mere electrions should, in collisions of atoms, sometimes be shot out with velocities considerably exceeding the velocity of light. Within an atom they would need to have far greater velocities than the velocity of light to fulfil the demands of your first theory for radium. I never however could satisfy myself as to the validity of that theory: and I have been supposing that you have not decidedly abided in it.[53]

Assuring Kelvin that he did abide in the theory, Thomson also assured Kelvin that it was unnecessary for corpuscles to exceed the speed of light in order to radiate energy. They only had to be accelerating. 'This does not require the velocity to be greater than that of light. In fact the results I have quoted would probably not hold in that case.'[54] Adding other objections, Kelvin's reply also showed that Thomson had evidently mistaken the nature of the first one.

> It seems to me that the supposed store of fifteen hundred years' energy in a radium atom, in the form of kinetic energy of electrions, or of "corpuscles", both vitreous and resinous, moving within it, would require velocities far exceeding the velocity of light.
>
> It seems to me a fatal objection to that theory that it implies each radium atom, with the "vast number of corpuscles" moving in it[s]

interior, should be predestined to come to the instability which causes them to be shot out at a predetermined time. When did the history of this state of affairs begin for each atom?, and what external influence gave it its moving corpuscles? What would be the difference, between radium atoms in a piece of radium-bromide, of the performance of those of the atoms which are nearly ripe for explosion, and those which have the prospect of several thousand years of stable diminishing motions before explosion?[55]

In other words, Thomson's corpuscles must have initially had a speed far greater than that of light, in order for them to have been able to slow down (as they radiated energy) for hundreds of years and still have speeds near that of light. But this situation was not physically possible. Kelvin had estimated—and in his support he cited the work of the prominent Maxwellian, Oliver Heaviside—that while it was possible for a material particle to travel faster than light, it 'requires the application of a continual pull forward to keep it moving uniformly at any constant velocity exceeding the velocity of light'.[56] The velocity of light was not an unsurpassable barrier to the particle's speed, but it was the threshold at which Newton's first law of motion ceased to be valid. Hence, with no appropriate force acting upon them, Thomson's corpuscles would slow down faster than he had thought and therefore could not spend centuries orbiting within the atom at speeds greater than that of light. Thomson and Rutherford, on the other hand, accepted much of the current 'electromagnetic' view of nature. According to it, an electron's mass was sufficiently electromagnetic that it increased asymptotically towards infinity as the electron's speed approached the speed of light.[57] Hence, even at speeds below that of light, the electron could attain an unlimitedly large mass and, therefore, kinetic energy. It could radiate energy for centuries and still be travelling near, though not at or above, the speed of light. Kelvin, however, as early as 1902 thought that beta-particles could travel faster than light, and he credited Heaviside with correcting 'an erroneous hypothesis, that no force however great could give an atom a velocity equal to the velocity of light, which has been somewhat extensively adopted within the last ten years in speculations and reckonings regarding radioactivity'.[58] Moreover, the Boscovichian approach gave Kelvin a force great enough to produce high-speed beta-particles, including, if necessary, ones exceeding the speed of light. Consequently, the Kelvin–Rutherford dispute turned in part on the contested point of the physical significance of the speed of light. Whereas Rutherford regarded intense acceleration as a fatal physical flaw of the static atomic model, Kelvin regarded excessive speed as a fatal flaw of the kinetic model.

In addition, Kelvin's theories of external energy and reversible atomic processes seemed more in accord with the behaviour of radio-active substances than was the theory of internal energy and dis-integrating atoms. If radioactive atoms came into being simultaneously (at the Creation) and if their eventual instability resulted only from an internal winding down, then why did they not all reach instability together? If they were not formed simultaneously, then what physical causes formed them during the course of time with their high-speed electrions? If they could be formed by natural processes, then was not radioactivity reversible? Did not the lack of simultaneity in disinte-gration therefore imply both external energy sources and reversible processes, in contradistinction to the disintegration theory? One support of Kelvin's views was James Dewar's report to him in 1903 that very cold temperatures considerably lowered radium's rate of radioactivity.[59]

Since Kelvin obviously thought he was able to absorb experimental facts of radioactivity into his programme of research better than Thomson and Rutherford could into theirs, it is hardly surprising that he continued to deny that his theories of the earth and sun had been overthrown. Asked about this by a former student early in 1906, Kelvin answered:

> The Gravitational theory is amply sufficient to account for the heat of both bodies, and it seems almost infinitely improbable that Radium adds practically to their energy for emission of heat and light. It may be indeed more probable that the energy of Radium may have come originally in connection with the excessively high temperatures, which we know to have been produced and to be at present being produced by Gravitational action throughout the Universe.[60]

Referring his correspondent to his earlier writings on the subject, Kelvin concluded that 'most of this evidence would not be seriously affected even if Radium concurred appreciably with Gravitation in producing the heat and light which we have at present in the Universe'.[61] He repeated his claims in his letter to *The Times* later the same year, 'protesting against the hypothesis that the heat of the sun or earth, or other bodies in the universe is due to radium'.[62]

Let us summarise Kelvin's position. He evidently regarded his old arguments as still sound. They did account for the amount of heat and light currently being emitted from the sun. Radium, unlike helium, was not spectroscopically identifiable in the sun. His tidal retardation theory for the earth's age still held. Hence, from this secure vantage point, just as other claims of the disintegration theorists looked shaky, so also did their speculations about the extended ages of the sun and earth. Specifically, not only could they not find radium in the sun but in regard to the earth their claim that radium behaved constantly was problematic. Dewar's 1903 letter told Kelvin otherwise, and, moreover, the extreme conditions sometimes obtaining in nature might be difficult to duplicate in the laboratory.[63] Kelvin thought it possible that the earth's early high temperatures might have influenced radium, in effect 'loading' radium atoms with electrions of great potential energy. Furthermore, 'a molecule of radium embedded in the earth's crust, under enormous pressures, probably has its constituent atoms safely protected against the explosive flyings asunder by which they produce the heating effects discovered in our laboratories'.[64] Hence, even if radioactive decay series had been unambiguously worked out, Kelvin could still propose that the radium now in existence had continually existed since the extreme conditions of the early earth had driven lead and helium together.[65] Because the essential constituent of radium remained intact, it was always physically capable of absorbing and emitting large amounts of energy. In the history of the earth, however, absorption seemed primarily confined to the early period of extreme conditions, emission to later periods. But the overall amount of energy involved was still about the same as Kelvin had earlier assumed. Now, instead of simply being radiated away from the earth, some of the initial energy had gone into loading radium atoms. Though Kelvin did not make the point, the apparent consequence of this view would have been an earth cooling somewhat more quickly to begin with, more slowly later. This, in turn, could possibly make the *habitable* period of the earth somewhat longer than Kelvin had previously calculated. Even here, however, the tidal retardation theory, which depended on the time at which the earth's surface solidified, would preserve Kelvin's main claim that the earth had been habitable for only a limited period. Clearly, from Kelvin's viewpoint, the uncertain speculations of the disintegrationists did not force him to abandon his long-established theory of a young earth.

There were, however, obvious experimental results which Kelvin ignored in his publications, and a balanced treatment of his viewpoint requires their discussion. The results strengthen as well as weaken Kelvin's case, leaving his main thrust, I should judge, untouched.

Rutherford, for example, reported experimental determination of a link between beta- and gamma-radiation.

> The intensity of the γ rays is always proportional to the rate of expulsion of β particles, and the result indicates that there is a close connection between the β and γ rays. Such a result is to be expected if the β particle is the parent of the γ ray, for the expulsion of each electron

from radium will give rise to a narrow spherical pulse travelling from the point of disturbance with the velocity of light.[66]

Had Kelvin noted this connection, however, he could reasonably have claimed that the intensely accelerated electrion in his atom was a likelier parent to a high-energy gamma-ray than was the only slightly accelerated electron in Rutherford's.

On the other hand, both Kelvin's suggestion that an alpha-particle was a recoiling radium atom and his claim that alpha-particles produce less energy than beta-particles clearly contradicted well established findings as reported by Thomson and Rutherford. How would these results have affected Kelvin's conclusions? It would seem that, at most, the fact of the alpha-particles' greater energy might have called into question Kelvin's decision to concentrate on finding the process by which beta-particles originate, but would not have detracted at all from his theory of static electrions and large repulsive forces. No longer being able to equate alpha-particles with recoiling atoms would have been more serious, leaving Kelvin with no explanation of the origin of alpha-particles.[67]

In this case it would seem natural for Kelvin to have done as Rutherford did and envisage similar origins for both alpha- and beta-particles. Rutherford wrote in 1905 that 'since the alpha particle is atomic in size, it is natural to suppose that the atoms of the radioactive elements consist not only of the electrons in motion, but also of positively charged particles whose mass is about the same as that of the hydrogen or helium atom'.[68] The parallel move for Kelvin would have been either somehow to adapt the force curve for beta-particles to alpha-particles or, more probably, to postulate a second force curve within the atom to govern alpha-particles. A strong repulsive force near the atom's centre would have explained the ejection of alpha-particles and might have fitted nicely with the later experimental results that led Rutherford to postulate a nuclear atom. At worst, as a cautious realist long used to formulating partial models in order to gain some insight into intractable physical processes, Kelvin could have admitted that his atomic model offered a clue only for the origin of beta-particles, but that it was superior to the Thomson–Rutherford atom in that regard.

Though such speculations on speculations can hardly be definitive, they at least indicate that these ignored experimental results did not unambiguously undermine Kelvin's whole research programme. Confident that Rutherford's conclusion about the earth's age was mistaken, Kelvin—from the perspective of his own physical research programme—questioned the series of assumptions which led to the conclusion. In doing so he highlighted some of the uncertainties in current physical theory. And he himself deemed it unnecessary to concede that radioactivity was an irreversible atomic disintegration.

A Victorian natural philosopher

Rather than a discontinuity between a young, alert William Thomson and an aged Lord Kelvin, I should like to emphasise continuity between his earlier work and that on radioactivity.

First, one should not automatically dismiss Kelvin's publications on radioactivity just because of the contrast between his few short papers and the mass of articles and books mushrooming from his opponents. Some of Kelvin's most seminal papers had been his shortest. Brief papers of 1842, 1847 and 1856 had crucially influenced Maxwell's physical thinking. 'The "Doctrine of Uniformity" in Geology Briefly Refuted' was two pages long. His vortex atom paper of 1867 ran to only ten pages.

Second, Kelvin was not simply unalterable in his opinions, incapable of accepting or creating new ideas. Agreeing in 1904 that gamma-rays were like x-rays, he corrected his earlier suggestion that gamma-rays might be a radioactive gas.[69] He changed his mind about the capacity of atoms to store energy for long periods. The sphere of positive electricity in J J Thomson's atom was, after all, Kelvin's innovation. By any standard, his Boscovichian atom—whether right or wrong— was an imaginative solution to a particular problem. As we saw in Chapter 7, he was even willing to assume that two chunks of matter could simultaneously occupy the same space.

Third, Kelvin was certainly not fully abreast of all the experimental results, in spite of his attention to Rutherford's and J J Thomson's books. Age may have been a factor. However, William Thomson had not been particularly noted for systematically keeping up with the literature. Often, he would write Stokes to ask what was known in a specific area. He was unaware of Helmholtz's 1847 paper on the conservation of energy until 1852.[70] Merton has counted how often Thomson unknowingly duplicated the work of others, a frequency attributed by Brush to Thomson's haphazard reading of the literature. Brush makes this point especially for a very young Thomson, less than 30 years old.[71]

Linked to the third point is the fourth. As before, in pondering radioactivity, Kelvin was mainly following the internal logic of his own physical insight, often arriving at conclusions regarded dubiously by others. We need only recall that for years his second law of thermodynamics gathered few followers. But we could well add that Stokes and Tait decisively rejected his concept of the aerial ether and that late-Victorian Maxwellians grew frustrated with his elastic-solid ether. His admirer and biographer, S P Thompson, marvelled at how 'impenetrable' he could sometimes be to the views of others.[72] One is reminded of Cannon's suggestion as to the two secrets of Darwin's greatness: 'One is Darwin's notorious habit of jumping to conclusions without adequate evidence. . . . The other is that of stubbornly maintaining his theories regardless of the valid arguments and evidence that could be brought against them.'[73]

Kelvin's advanced age was significant, however. Not only Rutherford but also his chief allies were products of an increasingly modern and secular era. Even Rutherford's mentor, J J Thomson, took his Cambridge degree twenty years *after* publication of *Essays and Reviews* and *Origin of Species*. Theological considerations, constituting much of the justificatory context of Kelvin's physical theories, were ignored by Rutherford, 'a man who was known to be completely indifferent to religion of any type'.[74] Citing the favourable implications for evolutionary theory, Rutherford seemed to *welcome* the demise of the young-earth theory, a demise that would have been most unwelcome to Kelvin. Kelvin's advanced age, therefore, signifies that his conflict with Rutherford was largely a *generational* dispute. Not merely slight thoughts of an old man, Kelvin's ideas were those of a profoundly speculative *Victorian* natural philosopher.

1   A S Eve 1939 *Rutherford: Being the Life and Letters of the Rt. Hon. Lord Rutherford, O.M.* (New York: Macmillan) p. 29.
2   *Ibid.* p. 107.
3   *Ibid.* pp. 108–9.
4   Lawrence Badash 1975 'Rutherford, Ernest' *Dictionary of Scientific Biography* vol. XII (New York: Scribner) p. 28. See Badash 1966 'How the "Newer Alchemy" Was Received' *Scientific American* **215** (August) 88–95.
5   Alex Keller 1983 *The Infancy of Atomic Physics: Hercules in His Cradle* (Oxford: Clarendon) p. 138. For similar statements, see Norman Feather *Lord Rutherford* (New York: Crane, Russak, 1973 reprint of 1940 edn) p. 94; and E N da C Andrade 1964 *Rutherford and the Nature of the Atom* (Garden City: Doubleday) pp. 74–5.
6   Thompson, for example, merely lists Kelvin's papers on radioactivity in a footnote. (*Thompson* II 1083n.)
7   In writing this section I have been able to draw on many studies including: Alfred Romer 1960 *The Restless Atom* (Garden City: Doubleday); Thaddeus J Trenn 1977 *The Self-Splitting Atom: The History*

of the Rutherford–Soddy Collaboration (London: Taylor and Francis); Joe D Burchfield 1975 *Lord Kelvin and the Age of the Earth* (New York: Science History Publications) ch. 6; John L Heilbron 1967–8 'The Scattering of α and β Particles and Rutherford's Atom' *Arch. Hist. Exact Sci.* **4** 247–307; Heilbron 1977 'Lectures on the History of Atomic Physics 1900–1922' in C Weiner (ed) *History of Twentieth Century Physics* (New York and London: Academic) pp. 40–108; and Heilbron 1981 'Rutherford-Bohr Atom' *Am. J. Phys.* **49** 223–31.

8   J J Thomson 1907 *The Corpuscular Theory of Matter* (London: Archibald Constable) p. vi. Rutherford wrote similarly, for example, in *Radio-Activity* (Cambridge: Cambridge University Press, 1904) p. vii, and *Radioactive Transformations* (New York: Scribner, 1906) p. 256.

9   J J Thomson 1904 *Electricity and Matter* (New Haven: Yale University Press) p. 96.

10 *Ibid.* p. 144.

11 Thomson *Corpuscular Theory of Matter* (note 8) p. vi.

12 Thomson *Electricity and Matter* (note 9) pp. 156–7.

13 Rutherford *Radio-Activity* (note 8) p. 126. See also pp. 337–42.

14 Ernest Rutherford 'The Radiation and Emanation of Radium. II' *The Collected Papers of Lord Rutherford of Nelson* 3 vols (London: Allen and Unwin, 1962–5) I 657. See also Rutherford, *Radio-Activity* (note 8) pp. 342–6.

15 See Burchfield *Kelvin and the Age of the Earth* (note 7) pp. 171–9.

16 Thomson *Corpuscular Theory of Matter* (note 8) p. 121.

17 H A Lorentz to Kelvin, 2 December 1901 *CUL Kelvin Collection* L106.

18 Kelvin *Baltimore Lectures* (1904) p. 123. The passage dates from 1884, and the index entry for it, dating from 1904, reads: 'Boscovich wrongly judged obsolete.'

19 Kelvin 'Molecular Constitution of Matter' *MPP* III 398.

20 Kelvin 'Extract from Address of November 30, 1893' *PLA* II 539.

21 Kelvin's note written on Joseph Larmor to Kelvin, 20 August 1900 *CUL Kelvin Collection* L25. Larmor wrote: 'I still think I prefer generalized dynamics without a definite model, but with its possibilities *illustrated* by the potentialities of the vortex atom theory'. Kelvin underlined *potentialities* and wrote beneath it, '(zero)'.

22 *Baltimore Lectures* (1904) pp. 300–1.

23 Kelvin's research notebook *CUL Kelvin Collection* NB151.

24 Kelvin 'Contact-Electricity and Electrolysis According to Father Boscovich' *MPP* VI 146. Omitted in the reprinted version, the diagram is from *Nature* **56** (27 May 1897) 84. The Royal Institution address is 'Contact Electricity of Metals' *MPP* VI 110–45. It discusses the controversies, for which see also *Thompson* I 401; II 996–7.

25 Kelvin 'Contact Electricity and Electrolysis According to Father Boscovich' (note 24) p. 147.

26 Kelvin 1902 'Aepinus Atomized' *Phil. Mag.* **3** (March) 263. The article is summarised in *Thompson* II 1078–80.

27 Kelvin 'Aepinus Atomized' (note 26) pp. 264–5.

28 *Ibid.* p. 267.

29   *Ibid.* p. 270.

30   *Ibid.* p. 273.

31   Kelvin 'Contribution to Discussion on the Nature of the Emanations from Radium' *MPP* VI 207–8.

32   *Ibid.* p. 208.

33   *Ibid.* p. 209. Just before the British Association meeting Kelvin had written to Dewar: 'I cannot think how Rutherford and others can complacently attribute the emission of heat by radium to energy stored in the substance.' (Kelvin to James Dewar, 29 August 1903 *Royal Institution Dewar Correspondence.*)

34   An entry in his research notebook for 14 August 1904 says he had been reading about alpha-rays in Rutherford's book. On 14 September he was reading Thomson's *Electricity and Matter* (*CUL Kelvin Collection* NB168.)

35   Kelvin's reply to Armstrong's letter proposing the quinine analogy. Armstrong to Kelvin, 4 September 1906, and Kelvin to Armstrong, 13 September 1906 *CUL Kelvin Collection* A78, A79.

36   Kelvin 'Plan of a Combination of Atoms Having the Properties of Polonium or Radium' *MPP* VI 217.

37   *Ibid.* p. 220.

38   Kelvin's research notebook *CUL Kelvin Collection* NB171, and Kelvin 'Plan of an Atom to be Capable of Storing an Electrion with Enormous Energy for Radio-Activity' *MPP* VI 228.

39   *Ibid.* p. 228.

40   Kelvin 'An Attempt to Explain the Radioactivity of Radium' *MPP* VI 234.

41   Thomson *Corpuscular Theory of Matter* (note 8) p. 2.

42   *Ibid.* pp. 1, 160–1.

43   Kelvin 'Contributions to Discussion on the Nature of the Emanations from Radium' (note 31) p. 209.

44   Kelvin 'Plan of a Combination of Atoms Having the Properties of Polonium or Radium' (note 36) p. 222.

45   Rutherford *Radio-Activity* (note 8) p. vii.

46   J J Thomson 1936 *Recollections and Reflections* (London: G Bell and Sons) p. 420.

47   Kelvin to James Dewar, 23 August 1903 *Royal Institution Dewar Correspondence.*

48   Kelvin to *The Times* 5 August 1906, *The Times* (9 August 1906) 3.

49   Kelvin to *The Times* 16 August 1906, *The Times* (20 August 1906) 6. Kelvin corrected the number for helium atoms in 'The Recent Radium Controversy' *Nature* **74** (27 September 1906) 539.

50   The opponents were: R J Strutt 'Radium' *The Times* (15 August 1906) 15; Oliver Lodge 'Radium' *The Times* (15 August 1906); Strutt 'Radium' *The Times* (21 August 1906) 4; A S Eve 'Radium' *The Times* (28 August 1906) 6; Frederick Soddy 'Radium' *The Times* (31 August 1906) 6; Lodge 'Radium' *The Times* (4 September 1906) 6; Soddy 'The Recent Controversy on Radium' *Nature* **74** (20 September 1906) 516–18; and Rutherford 'The Recent Radium Controversy' *Nature* **74** (25 October

1906) 634–5. Rutherford's long quotation from *Radio-Activity* emphasised 'that this postulated helium compound is of a character entirely different from that of any other compound previously observed in chemistry'.

51   Frederick Soddy's report of a discussion with Kelvin in Soddy 'Radium' *The Times* (31 August 1906) 6. For accounts of the controversy see Burchfield *Lord Kelvin and the Age of the Earth* (note 7) p. 165; Eve *Rutherford* (note 1) pp. 140–2; and *Thompson* II 1190–1.

52   Harold A. Wilson 'Is Radium an Element?' *Nature* **70** (14 July 1904) 241. In addition to Rutherford's *Radio-Activity*, Wilson was reviewing F Soddy 1904 *Radio-Activity: An Elementary Treatise from the Standpoint of the Disintegration Theory* (London: The Electrician) and L A Levy and H G Willis 1904 *Radium* (London: Percival Marshall). Kelvin pasted into his research notebook a vastly more critical review of Soddy's book which spoke of Soddy's 'airy visions' and declared that 'science is disfigured at the present day by too much rash theorizing upon imperfect *data*'. ('Radio-Activity' *The Times Literary Supplement* (22 July 1904) 230; Kelvin's research notebook *CUL Kelvin Collection* NB168.)

53   Kelvin to J J Thomson, 8 November 1906 *CUL Kelvin Collection* T539.

54   J J Thomson to Kelvin, 11 November 1906 *CUL Kelvin Collection* T540.

55   Kelvin to J J Thomson, 12 November 1906 *CUL Kelvin Collection* T541. Printed in Lord Rayleigh (R J Strutt) 1942 *The Life of Sir J. J. Thomson* (Cambridge: Cambridge University Press) pp. 141–2. Rayleigh thought that Kelvin's criticism 'foreshadows the insuperable difficulty that has been found in a determinist theory of the breaking up of radioactive atoms'.

56   Kelvin 'On the Motions of Ether Produced by Collisions of Atoms or Molecules, Containing or Not Containing Electrions' *MPP* VI 238. This 1907 paper referred back to his 1900 lecture 'Nineteenth-Century Clouds over the Dynamical Theory of Heat and Light' published as Appendix B in *Baltimore Lectures* (1904) §§4–7.

57   Rutherford *Radio-Activity* 2nd edn (Cambridge: Cambridge University Press, 1905) pp. 128–30, and Thomson *Electricity and Matter* (note 9) pp. 36–52. See Russell McCormmach 1970 'H. A. Lorentz and the Electromagnetic View of Nature' *Isis* **61** 459–97.

58   Kelvin 'On the Motions of Ether Produced by Collisions of Atoms or Molecules' p. 238. The 1902 paper is Kelvin 'Becquerel Rays and Radio-Activity' *MPP* VI 205.

59   Dewar wrote to Kelvin: 'The measurements I made of heat evolution of Radium Bromide per gram in Liquid Oxygen came out 38 gram units per hour and in Liquid Hydrogen the value came out 57 gram units per hour.' (Dewar to Kelvin, 26 August 1903 *CUL Kelvin Collection* D89.) Kelvin reported the lower of these values along with Curie's of 90 in his paper to the British Association meeting in 1903. His point there was that even Dewar's low value required more energy than an atom could contain. ('Contributions to Discussions on the Nature of the Emanations from Radium' (note 31) p. 208.)

60   Kelvin to James Orr, 29 January 1906 *CUL Kelvin Collection* O14.

61 *Ibid.*

62 Kelvin to *The Times* 5 August 1906 (note 48).

63 It should be noted that, in the case of the sun, Rutherford and his supporters also allowed for the rate of radioactive disintegration to vary. As Rutherford put it, 'It is not improbable that, at the enormous temperature of the sun, the breaking up of the elements into simpler forms may be taking place at a more rapid rate than on the earth.' (*Radio-Activity* (1904) p. 344. See Burchfield *Kelvin and the Age of the Earth* (note 7) pp. 167–8.) According to Kelvin's views, *if* radioactivity were present in the sun the high temperatures would presumably not only have increased the rate of radioactivity but also supplied enough energy to make the *reverse* process possible, thus complicating estimation of the overall effect.

64 Kelvin to *The Times* 22 August 1906, *The Times* (24 August 1906) 6.

65 *Ibid.* Having to admit that some, or even all, of the radium was derived from uranium would not, of course, have forced him to alter his basic claim.

66 Rutherford *Radio-Activity* (1905) p. 185.

67 The atoms did recoil, but they were not alpha-particles. See Thaddeus J Trenn 1975 'Rutherford and Recoil Atoms: The Metamorphosis and Success of a Once Stillborn Theory' *Hist. Stud. Phys. Sci.* **6** 513–47. Kelvin evidently did not connect the emission of alpha-particles with the production of helium, possibly because the letter informing him of the experimental verification of helium production mentioned nothing about alpha-particles. (William Ramsay to Kelvin, 30 August 1903 *CUL Kelvin Collection* R15.)

68 Rutherford *Radio-Activity* (1905) pp. 157–8.

69 Kelvin stated this in a note dated 23 June 1904 which he inserted in 'Contributions to Discussion on the Nature of the Emanations from Radium' (note 43) p. 206.

70 Thomson 'On the Dynamical Theory of Heat, with Numerical Results Deduced from Mr. Joule's Equivalent of a Thermal Unit, and M. Regnault's Observations on Steam' *MPP* I 182–3n.

71 Robert K Merton 1961 'Singletons and Multiples in Scientific Discovery: A Chapter in the Sociology of Science' *Proc. Am. Phil. Soc.* **105** 484–5, and Stephen G Brush 1976 *The Kind of Motion We Call Heat: A History of the Kinetic Theory of Gases in the 19th Century* 2 vols. (Amsterdam: North-Holland) II 331–2.

72 *Thompson* II 997.

73 Walter F Cannon 1961 'The Bases of Darwin's Achievement: A Revaluation' *Victorian Studies* **5** 134.

74 David Wilson 1983 *Rutherford: Simple Genius* (London: Hodder and Stoughton) p. 596.

Index